IMAGES
of Aviation

ST. LOUIS AVIATION

ON THE COVER: One of the three Lindbergh Ryans made especially for the 1957 motion picture production of *The Spirit of St. Louis* is being flown past the recently completed Gateway Arch Monument and Grounds in 1965.

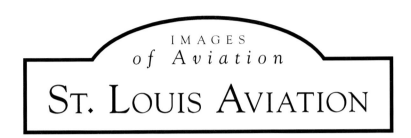

IMAGES
of Aviation

ST. LOUIS AVIATION

Jeremy R. C. Cox
for the St. Louis Air and Space Museum

ARCADIA
PUBLISHING

Published by Arcadia Publishing
Charleston, South Carolina

Printed in the United States of America

Library of Congress Control Number: 2009943883

For all general information, please contact Arcadia Publishing:
Telephone 843-853-2070
Fax 843-853-0044
E-mail sales@arcadiapublishing.com
For customer service and orders:
Toll-Free 1-888-313-2665

Visit us on the Internet at www.arcadiapublishing.com

CONTENTS

ACKNOWLEDGMENTS

This book would not have been possible without the kind and gracious assistance and encouragement provided to me first by my dear wife, Deb, and also by my fellow aviation history enthusiasts at the Greater St. Louis Air and Space Museum. I would like to specifically thank the following people: Jack Abercrombie, curator; Mark Nankivil, chief archivist; David Sneddon, museum librarian; and board members Chubb Wheeler, Rick Rehg, Bob Crandall, Bob Kraemer, Joe Gutknecht, Jean Murry, Jan Pocock, Trent Duff, and Jeff Reich. I would also like to give special mention to Margret Koch of the Missouri History Museum; Bob McDaniel, the airport director at St. Louis Downtown Airport; and his assistant Wendi Sellers—all three of whom have been great friends to the Greater St. Louis Air and Space Museum over the years. Also, it would be wrong to not mention Gregg Merinack of the St. Louis Science Center, Ed Sutorius, Dr. Fred Roos, Sean Goding, Diane Earhart, the Saugets, Bill Florich, and Erik Lindbergh. If it were not for all of you, the museum would not have lived.

Every image in this book is from the archives that have been donated over the years and are now owned and maintained by the Greater St. Louis Air and Space Museum.

—Jeremy R. C. Cox
President of the Greater St. Louis Air and Space Museum

INTRODUCTION

The Greater St. Louis Air and Space Museum is the St. Louis Air and Space Museum and was incorporated in July 1982 to preserve the rich aeronautical heritage that is centered on the Greater St. Louis region. The museum is a not-for-profit Missouri corporation with federal tax-exemption status under tax code 501(c)(3). The Greater St. Louis Air and Space Museum continues as a Missouri charitable corporation operating in Illinois. The museum is located within the historic Hangar 2 at the St. Louis Downtown Airport (KCPS) in Cahokia/Sauget, Illinois.

The fundamental purpose of the Greater St. Louis Air and Space Museum is education. The region's substantial contributions to the development of aerospace will be preserved and displayed for the enjoyment and enlightenment of future generations. Besides the static exhibits, an active education program will be conducted on-site and in area schools.

The mission of the museum is to serve the public, which is done through the following: preserving and displaying of historic air and space craft and artifacts and providing educational programs to foster the spirit of flight in today's youth and in future generations.

The author's proceeds from the sale of this book go to support the museum.

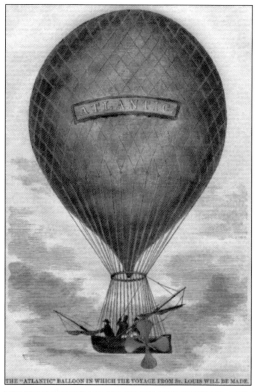

THE "ATLANTIC" BALLOON IN WHICH THE VOYAGE FROM St. LOUIS WILL BE MADE.

John Wise's *Atlantic* gas balloon is shown in flight in 1859. The basket is an equipped rowing skiff in case Wise and his crew had to put down in the ocean. Wise and fellow aviators were not the first aeronauts to fly here though. In fact, the rich aviation history of St. Louis began on Tuesday, May 17, 1836, when the travelling hydrogen-gas-balloon pilot, Richard Clayton, made a 6-mile demonstration flight that started from the intersection of Fourth and Market Streets in St. Louis. Originally from England, Clayton immigrated to Cincinnati as a clock maker, silversmith, and budding aeronaut. Once he had perfected a reliable balloon design, he travelled extensively throughout the Midwest giving "paid-admission only" demonstrations of his flying machine.

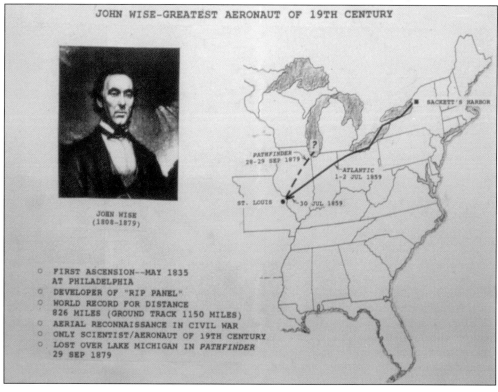

JOHN WISE—GREATEST AERONAUT OF 19TH CENTURY

JOHN WISE
(1808–1879)

○ FIRST ASCENSION—MAY 1835
  AT PHILADELPHIA
○ DEVELOPER OF "RIP PANEL"
○ WORLD RECORD FOR DISTANCE
  826 MILES (GROUND TRACK 1150 MILES)
○ AERIAL RECONNAISSANCE IN CIVIL WAR
○ ONLY SCIENTIST/AERONAUT OF 19TH CENTURY
○ LOST OVER LAKE MICHIGAN IN *PATHFINDER*
  29 SEP 1879

One

THE BEGINNING OF

FLIGHT IN ST. LOUIS

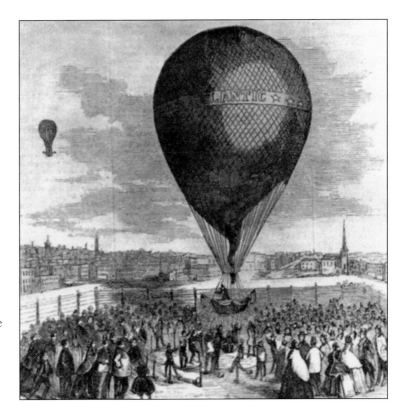

Pictured here in 1859, John Wise and his crew ascend from St. Louis in the 65,000-cubic-foot helium balloon, the *Atlantic*, on their attempt to cross the Atlantic Ocean.

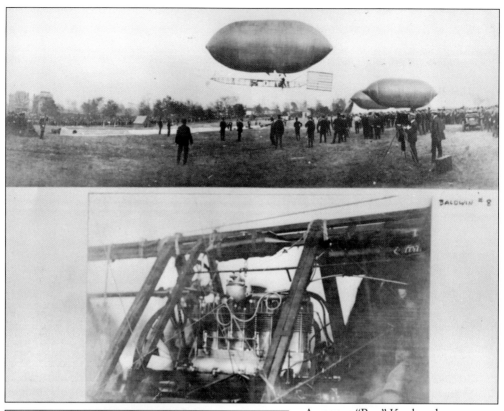

Augustus "Roy" Knabenshue demonstrates sustained and controlled flight of the *California Arrow* at the 1904 World's Fair in St. Louis. The airship was powered by the Curtiss motorcycle engine of Missouri-native friend Thomas Scott Baldwin.

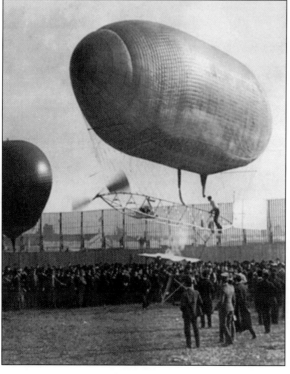

The *California Arrow* lifts off from the Forest Park Fairgrounds, which had been surrounded with barrier fences to prevent non-paying spectators from seeing too much of the action.

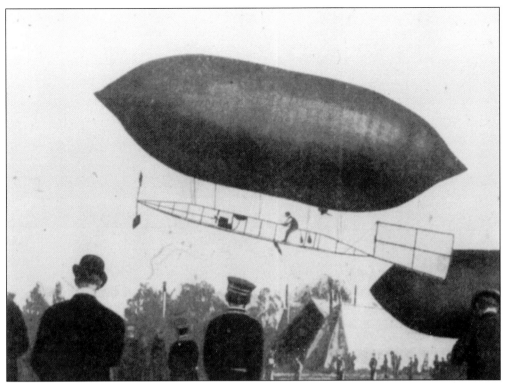

The Baldwin *California Arrow* climbs out for a circuit around the fairgrounds. Approximately 18,740,000 people attended the 1904 World's Fair that was held in St. Louis that year.

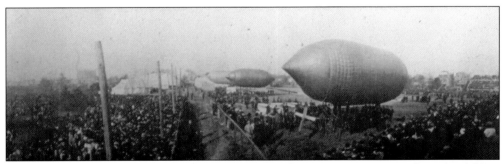

Formally named the Louisiana Purchase Exposition, this immense event ran from April 30 to December 1 and is notable for it being the first event anywhere that served ice cream cones, toasted raviolis, and demonstrated air-to-ground and ground-to-air wireless telegraphy. Dr. Pepper, peanut butter, iced tea, puffed wheat, hot dogs, and hamburgers were also all vended to a mass audience for the first time here.

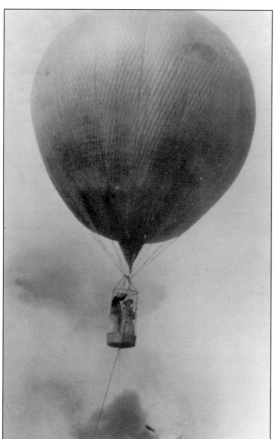

A tethered observation balloon at the 1904 World's Fair afforded fair-going passengers with an unparalleled view of the fair from high above. The popular song composed and written by Kerry Mills and Andrew Sterling, "Meet Me in St. Louis," was also courtesy of the seven-month-long extravaganza.

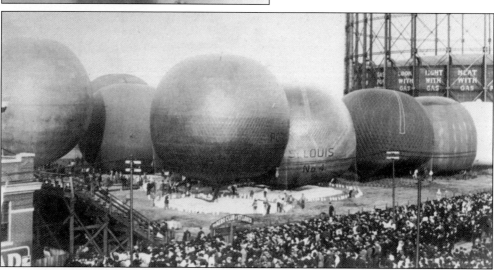

In 1907, the James Gordon Bennett Cup International Balloon Race, held in St. Louis for the first time, started from Forest Park. This event was sponsored by the Aero Club of St. Louis, which was founded by Lewis Dozier five months before the race. German balloonist Oscar Erbsloh, aided by famed meteorologist and balloonist H. H. Clayton, took the cup this year after flying a distance of 880 miles, landing in Asbury Park, New Jersey.

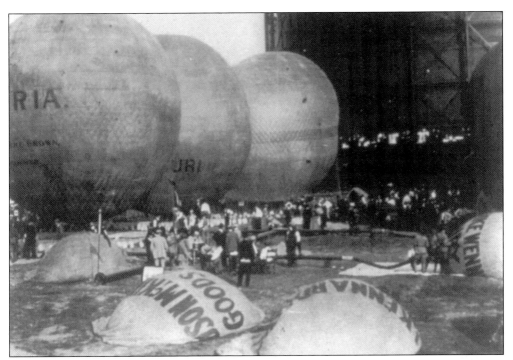

One of the city's massive "gasometers" inflates racing balloons before the cup race. This race, which still runs each year and is hosted by the country of residence of the winners from the previous year's race, was founded by James "Gordon" Bennett, the owner of the *New York Herald* newspaper. It was first run in Paris, the year before the St. Louis event. Its formal name is Coupe Aéronautique Gordon Bennett.

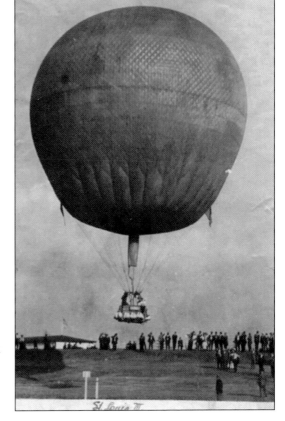

St. Louis aeronauts Albert Bond Lambert and Sylvester Louis Von Phul lift off in their balloon *St. Louis III* from Forest Park in 1909. While on their 685-mile flight that ended in a field in South Carolina, they subsequently set the world's balloon speed record of 44 miles per hour.

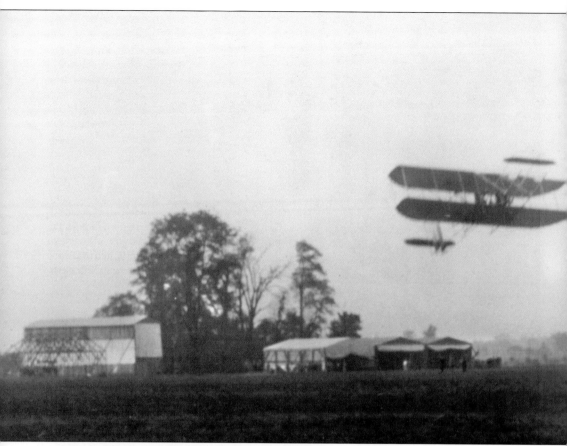

A Wright Flyer takes off at the newly established St. Louis airfield, Kinloch Park; the Aero Club of St. Louis opened this field. Kinloch Park was a horse racing track and fairground field that was chosen as the first permanent flying field in St. Louis after the trees at Forest Park proved too numerous for safe operation of the relatively un-maneuverable craft of the time. This newly dedicated flying site was relocated near the present intersection of North Hanley Road and Frost Avenue, northeast of Lambert-St. Louis International Airport. It was opened in May 1910 and ran continuously until the St. Louis Aero Club moved its operations to the 110-acre site (that would eventually become the present-day Missouri Lambert-St. Louis International Airport), located to the southwest, near Natural Bridge, Missouri.

Marion, Missouri, native Thomas Scott Baldwin poses in the pilot seat of his *Red Devil* aircraft at Kinloch Park on September 10, 1910, before flying down the Mississippi River to a landing near the Cahokia Ferry in Illinois. He flew under both the Eads and Merchant Bridges on his return flight path back to Kinloch Park. The *St. Louis Post Dispatch* sponsored the public aviation demonstration to show the city and the country that aviation was a practical reality.

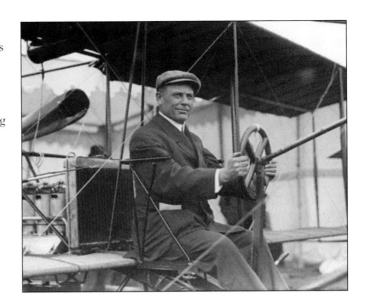

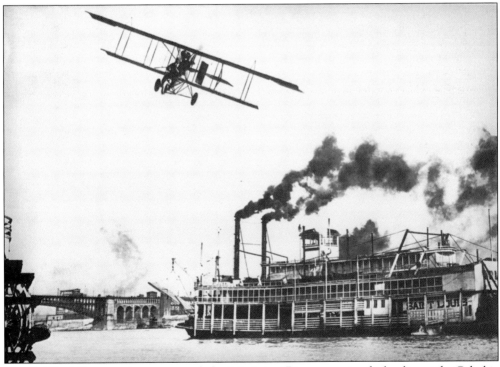

The *Red Devil* flies over the commerce-laden Mississippi River preparing for landing at the Cahokia Ferry. The Eads Bridge is seen in the background. Baldwin's design was the first to incorporate a largely steel-tube fuselage and wing structure. Powered by a 100-horsepower Hall-Scott V8 engine (this engine later became the basis of the design of the highly successful Liberty engine), the *Red Devil* got its name because it was covered with a rubberized red silk.

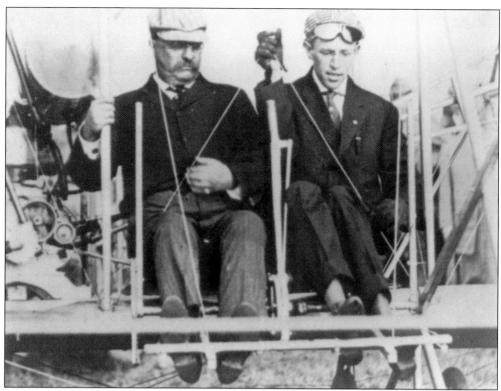

On Tuesday, October 11, 1910, Theodore Roosevelt became the first U.S. president to fly in an aircraft. In the photograph below, President Roosevelt and Pilot Hoxsey taxi away from the crowd before takeoff. Prior to this historic flight in front of 30,000 spectators, Hoxsey was heard saying, "Colonel Roosevelt, our birthdays fall on October 27th, and as a result, I feel that I have somewhat of a lien on your indulgence. I want you to take a short spin with me in the air. You can, with perfect safety, trust yourself in my hands." Colonel (President) Roosevelt pushed his hat back on his head and asked, "Now?" Hoxsey replied, "Yes. This is as good a time as we can ever get. It is calm and ideal flying-machine weather. Will you go?" The rest is history.

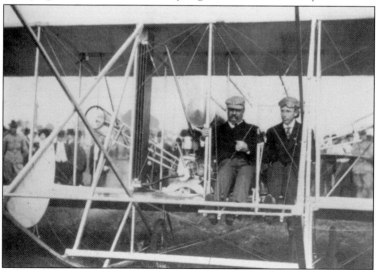

This image shows Roosevelt's flight just after landing. Travelling at a speed of 40 miles per hour, Arch Hoxsey piloted President Roosevelt on a 1-mile course over Kinloch Airfield at a height of 150 feet.

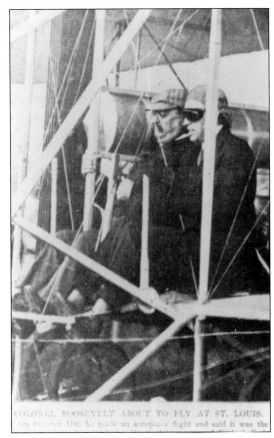

COLONEL ROOSEVELT ABOUT TO FLY AT ST. LOUIS.
On October 11th he made an aeroplane flight and said it was the

In 1918, Albert Berry (standing) made what is debatably the world's first parachute jump from a fixed-wing aircraft. Flown by Tony Jannus (seated) from Kinloch Airfield south to the Jefferson Barracks, Berry leapt from the aircraft at approximately 1,500 feet. Berry was observed climbing down to the wheel axle and hanging, trapeze-like, before he let go; his canopy deployed as he left the aircraft. Thomas Baldwin is credited with the first parachute jump in the United States; in 1887, Baldwin leapt from a balloon floating over San Francisco.

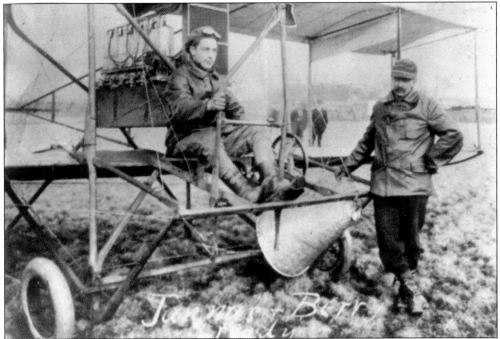

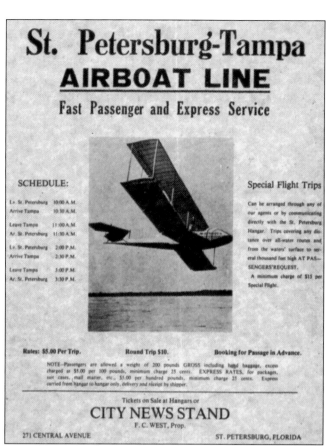

St. Petersburg-Tampa
AIRBOAT LINE
Fast Passenger and Express Service

SCHEDULE:

Lv. St. Petersburg 10:00 A.M.
Arrive Tampa 10:30 A.M.

Leave Tampa 11:00 A.M.
Ar. St. Petersburg 11:30 A.M.

Lv. St. Petersburg 2:00 P.M.
Arrive Tampa 2:30 P.M.

Leave Tampa 3:00 P.M.
Ar. St. Petersburg 3:30 P.M.

Special Flight Trips

Can be arranged through any of our agents or by communicating directly with the St. Petersburg Hangar. Trips covering any distance over all-water routes and from the waters' surface to several thousand feet high AT PASSENGERS' REQUEST.

A minimum charge of $15 per Special Flight.

Rates: $5.00 Per Trip.   **Round Trip $10.**   **Booking for Passage in Advance.**

NOTE—Passengers are allowed a weight of 200 pounds GROSS including hand baggage, excess charged at $5.00 per 100 pounds, minimum charge 25 cents. EXPRESS RATES, for packages, suit cases, mail matter, etc., $5.00 per hundred pounds, minimum charge 25 cents. Express carried from hangar to hangar only, delivery and receipt by shipper.

Tickets on Sale at Hangars or

CITY NEWS STAND
F. C. WEST, Prop.

271 CENTRAL AVENUE   ST. PETERSBURG, FLORIDA

On Thursday, January 1, 1914, Tony Jannus commanded the world's first scheduled airline flight. The route was flown from St. Petersburg to Tampa, Florida, in a Benoist Type 13 flying boat, which was designed and built in St. Louis by Missouri native Thomas W. Benoist.

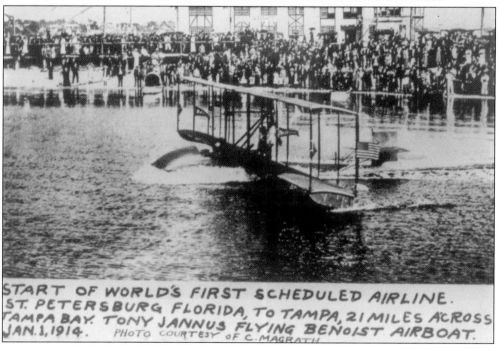

START OF WORLD'S FIRST SCHEDULED AIRLINE. ST. PETERSBURG FLORIDA, TO TAMPA, 21 MILES ACROSS TAMPA BAY. TONY JANNUS FLYING BENOIST AIRBOAT. JAN. 1, 1914. PHOTO COURTESY OF C. MAGRATH

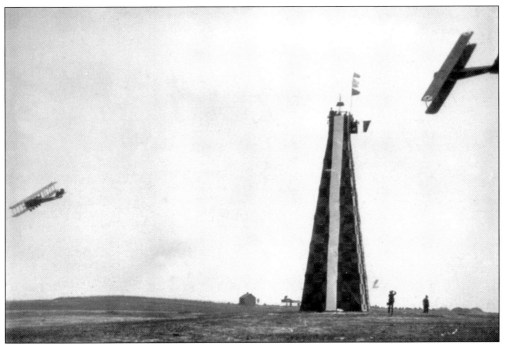

On October 1 and 2, 1923, St. Louis hosted the National Aero Congress of the National Aeronautic Association. From October 3 to 6, the International Air Races, known as the St. Louis Air Meet, were then held at Lambert Field.

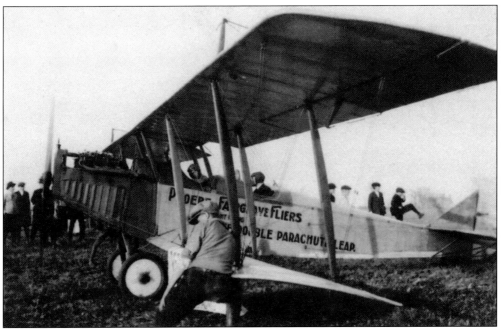

The St. Louis Air Meet parachutist demonstration team is pictured. They flew down from St. Paul, Minnesota, for the air meet. The parachutist is Phoebe Fairgrave Omlie, a female pilot who would later found the female pilot's association, the Ninety-Nines.

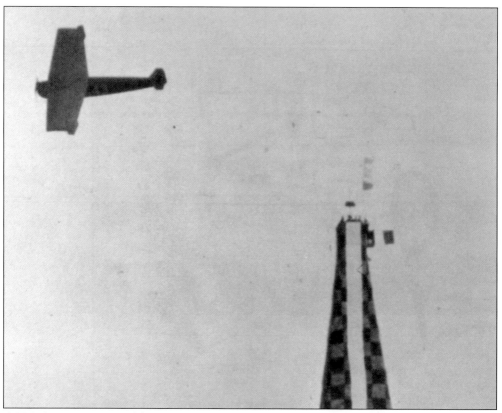

A U.S. Army Air Race Fokker T-2 aircraft flies past the finish-line pylon during the International Air Races. Earlier in May, the U.S. Army team set a coast-to-coast flight record in one of its T-2 aircraft, flying from New York to San Diego in 26 hours and 50 minutes.

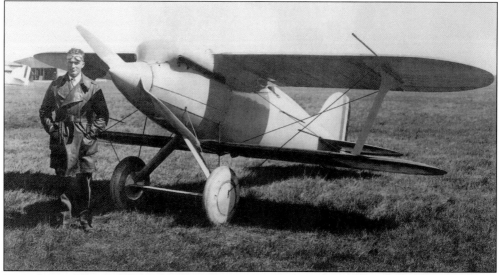

One of several Curtiss R2C1 U.S. Navy Racing aircraft that competed at the 1923 International Air Race in St. Louis, this model was credited with attaining 266 miles per hour. Lt. A. J. Williams, the winning pilot, poses with his aircraft.

20

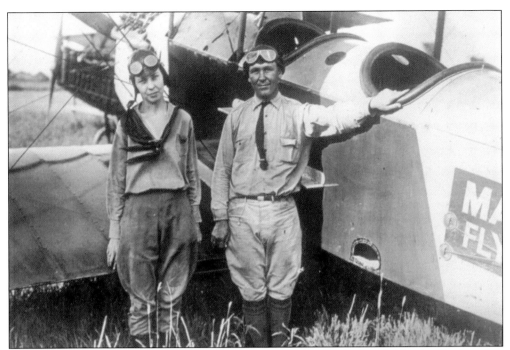

Young St. Louis native Marie Meyer is pictured here with her husband, Charles Fowler, after taking flying lessons at the Robertson Flying School at Lambert Field in 1923. She started a flying circus shortly after marrying. Marie and Charles staged daring displays of balance and courage in many death-defying feats with their aircraft. Wing walking and parachute jumping were two of Marie's fortes; Charles flew the aircraft.

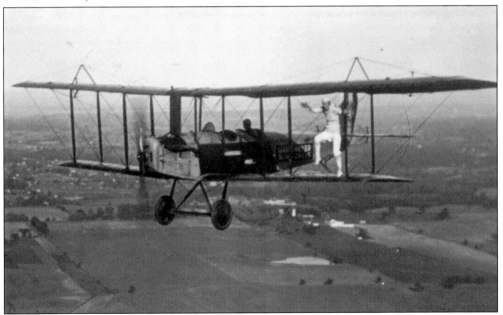

Marie Meyer walks on the wing with Charles at the controls of their 70-mile-per-hour Curtiss Standard J-1, ex–World War I training aircraft, which was powered by a 105-horsepower, V8 Curtiss OX-5 engine.

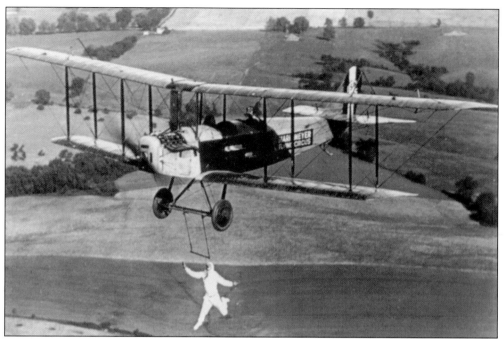

During Marie Meyer's aerial trapeze act, Meyer hung from the bar by her teeth. She neither wore a parachute nor had a safety harness.

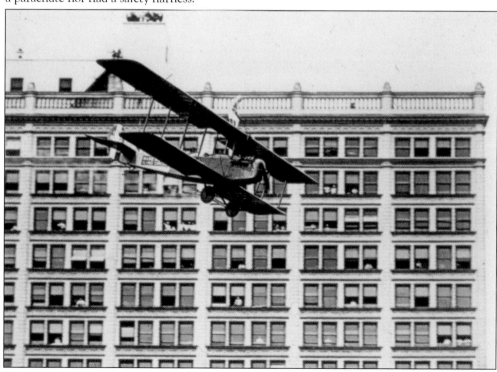

Marie Meyer is atop the wing of her biplane, and her husband, Charles, guides the aircraft down to perform a low-level pass through St. Louis. The Railway Exchange Building (later the Famous-Barr flagship store) can be seen in the background.

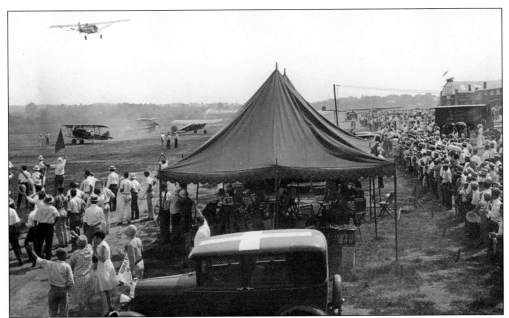

In 1929, from July 13 to 30, St. Louis pilots Dale Jackson and Forrest O'Brine took the world's endurance flight record away from pilots Reginald Robbins and James Kelly of Fort Worth, Texas. The St. Louis crew remained aloft in their *St. Louis Robin* for 420 hours and 21 minutes over Lambert Field.

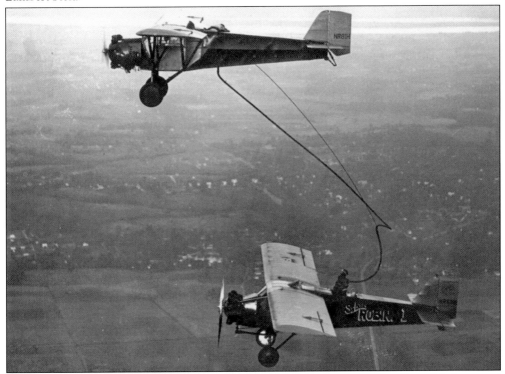

Missouri Air Guard pilot Maj. Ray Wassall flew the refueling aircraft that allowed Jackson and O'Brine to remain aloft without landing.

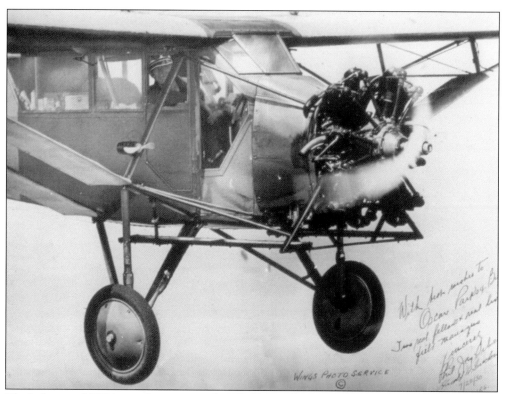

The Jackson and O'Brine endurance record was held for only 11 months. Brothers John and Kenneth Hunter, pilots from Chicago, took it from them after flying for 553 hours and 41 minutes. Not to be usurped, Jackson and O'Brine took to the skies of St. Louis again to re-win the record by staying aloft for 647 hours and 28 minutes. They had renamed their aircraft the *Greater St. Louis.*

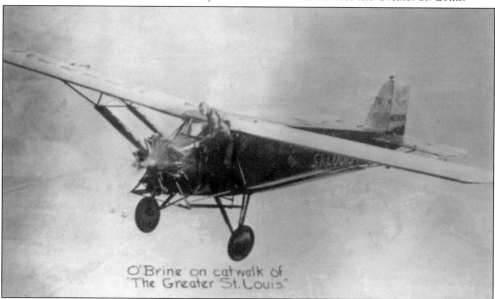

Forrest O'Brine performs maintenance on the *Robin's* engine. Both Forrest and Dale wore parachutes during their record-breaking endurance flights.

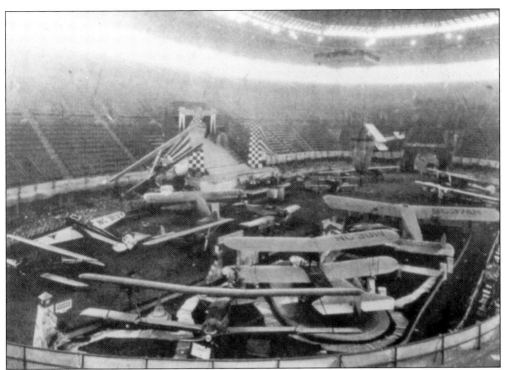

In February 1930, an International Aircraft Exposition was held within and around the St. Louis Arena (now demolished). About 170 aviation companies displayed products and services at this show, and a handful of the aircraft took off and landed in Forest Park. A world altitude record of 27,250 feet was set. Additionally, a record was set at Lambert Field for consecutive "outside" (negative "g") loops.

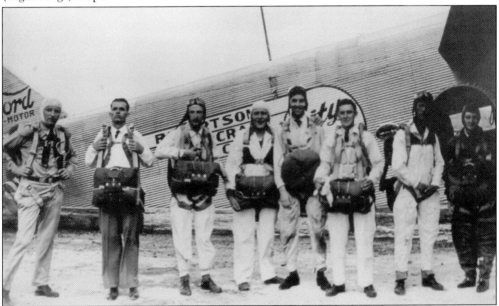

This 1930s parachute demonstration group poses for a photograph before stepping aboard its Ford Tri-motor, owned by Robertson Aircraft, and climbing to the "drop altitude."

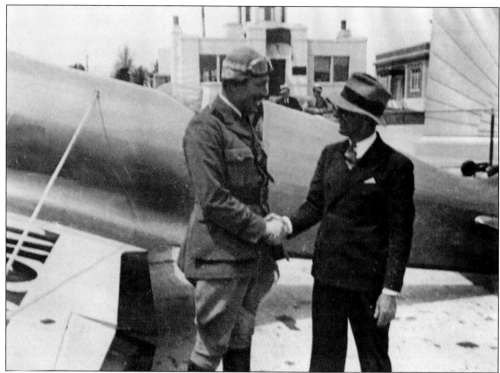

In May 1937, the St. Louis Air Races and International Aerobatic Competition was held at St. Louis-Lambert Municipal Airport. Here Col. Rosco Turner (left) is seen in attendance, but he did not compete in any of them. Colonel Turner is greeted by the St. Louis aviation pioneer Frank Robertson upon his arrival at Lambert.

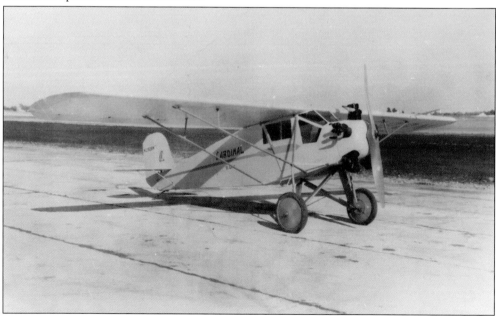

A St. Louis Cardinal is ready for delivery to its private buyer on the ramp at Lambert-St. Louis Municipal Airport. The Cardinal was powered by a 100-horsepower Kinner 5-cylinder radial engine.

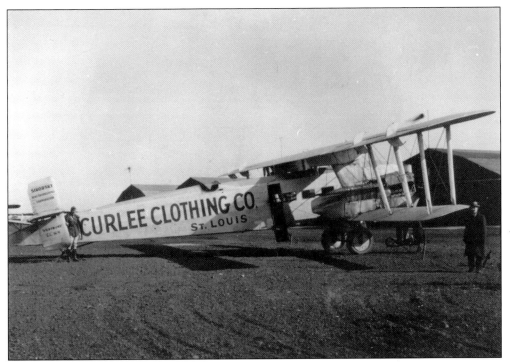

This 14-passenger behemoth, a Sikorsky S-29A transport, was sponsored by the Curlee Clothing Company of St. Louis and flown by Roscoe Turner. While making city tours around the country, the plane was used to promote the concept of airline transportation to the nation. Obviously the Curlee Clothing Company benefited greatly from the advertising as well.

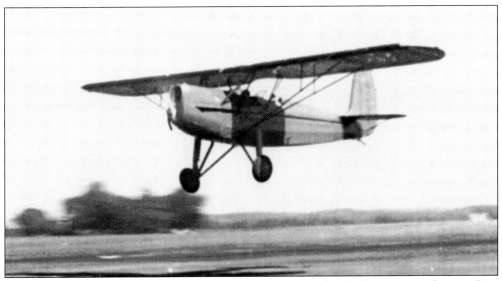

A privately owned Fairchild C7D makes a flyby in 1938 at Lakeside Airport near Granite City, Illinois (the town located across the McKinley Bridge from St. Louis).

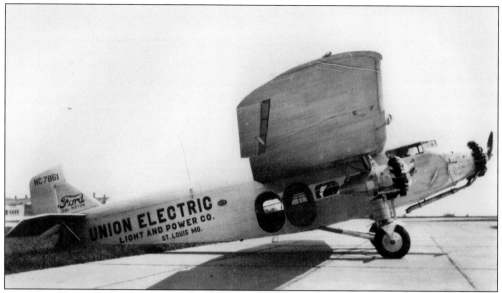

St. Louis–based Union Electric Corporation purchased a Ford 4-AT-B Tri-motor for survey work and as a corporate transport in 1930. The company based it at the Curtiss-Steinberg Airport in the same Curtiss-Wright hanger that the Greater St. Louis Air and Space Museum is housed in today. This aircraft can still be viewed during a visit to the U.S. Naval Museum in Pensacola, Florida, where it is on display in navy colors.

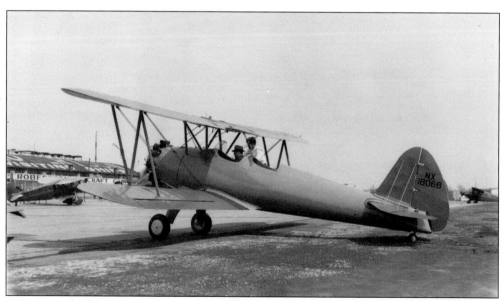

A St. Louis–built PT-1W primary trainer prepares for a test-flight hop at Lambert in 1939.

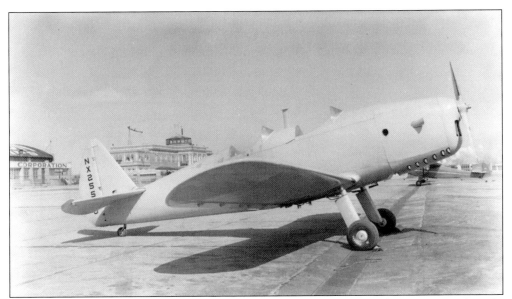

A St. Louis prototype all-metal monoplane trainer, the PT-LM-4, sits on the ramp at Lambert-St. Louis Municipal Airport in 1940.

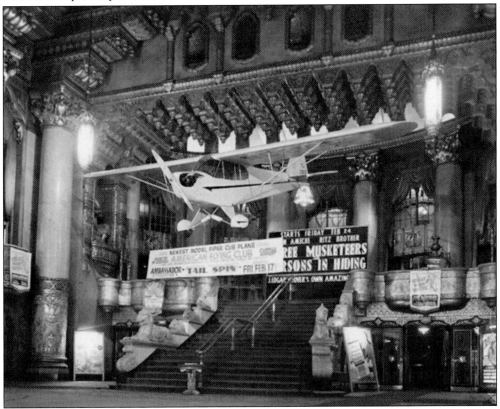

A 1940 Piper J3 Cub is on display in the lobby of the Fox Theater to advertise the American Flying Club of East St. Louis, which was based at the Curtiss-Steinberg Airport. By driving over the MacArthur Municipal Bridge, the airport was just south and west from downtown St. Louis.

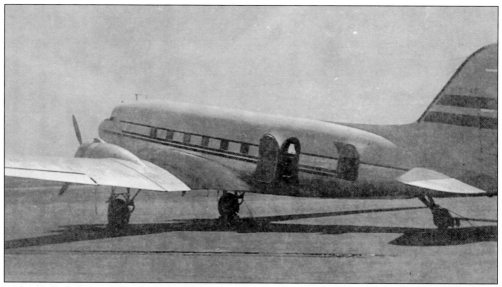

Museum board member Chubb Wheeler used to fly this aircraft, the corporate Douglas DC-3, which was owned by the St. Louis–based Peabody Coal Company.

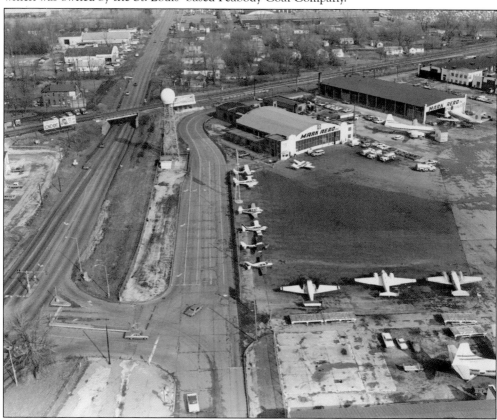

The Mark Aero facilities at Lambert-St. Louis International Airport are pictured in 1973. The entire aircraft-movement and hangar area on the northwest end of the airport was demolished soon after, with very little trace of it having ever existed remaining today.

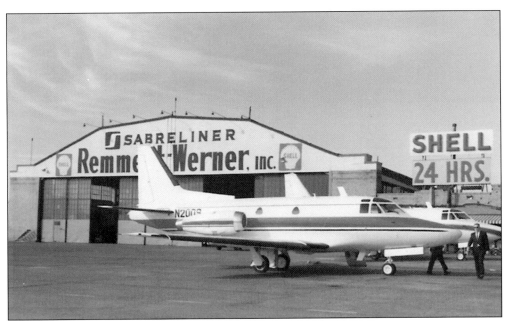

St. Louis–based Monsanto Corporation is still the world's oldest corporate jet–owning company. This image shows one of its first jets, a Rockwell Sabreliner 40, which was completed and sold to Monsanto by Remmert-Werner at Lambert. Remmert-Werner started its business-aircraft modification and completions business by first converting ex–World War II Douglas DC-3 aircraft into corporate/executive business aircraft.

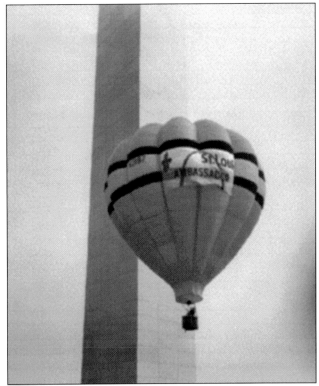

The late St. Louis native aviatrix Nikki Caplan, who founded the Great Forest Park Balloon Race, was given permission to fly her balloon through the Gateway Arch in 1973. She remains the only pilot to have received official sanction from the Federal Aviation Administration to fly through the Gateway Arch. The 630-foot monument, designed to recognize the "Westward expansion," was opened to the public on July 10, 1967. The arch was co-designed by architect Eero Saarinen and structural engineer Hannskarl Bandel.

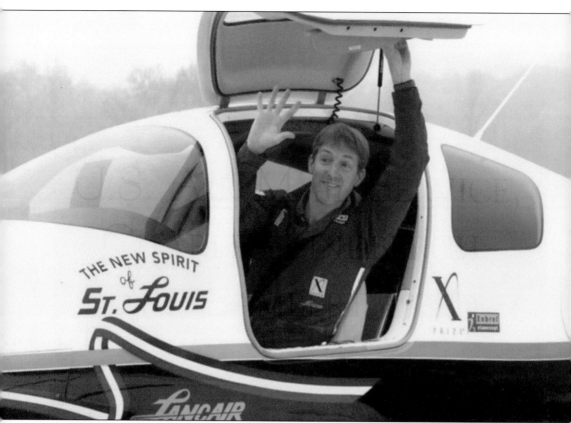

On May 1 and 2, 2002, with the assistance of Gregg Maryniak (the director of the James S. McDonnell Planetarium, the vice president for Energy and Aerospace of the St. Louis Science Center, and the corporate secretary and chairman of the Executive Committee of the Board of the X PRIZE Foundation), Erik Lindbergh retraced his grandfather Charles Augustus Lindbergh's solo flight from New York to Paris in a small single-engine aircraft named the *New Spirit of St. Louis*, which is a Lancair that was built in Oregon. The personal journey, documented by the History Channel, raised over $1 million for three charities, garnered half a billion media impressions for the XPRIZE Foundation, and prompted a call from the president of the United States for inspiring the country after the tragedy of September 11 the previous year. The X PRIZE Foundation is the brainchild of Dr. Peter H. Diamandis, who believes that focused and talented teams in pursuit of a prize and acclaim can change the world. Both the Orteig Prize and the flight of the original *Spirit of St. Louis* piloted by Charles Lindbergh inspired his foundation.

Two

U.S. Air Mail Service, Charles Lindbergh, and Lambert Field

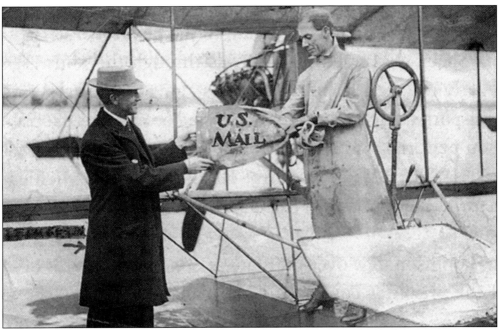

The St. Louis Aero Club sponsored a hydroplane (a landplane equipped with floats) demonstration in 1911. It was the first flight of this type of aircraft in St. Louis history. Hugh Robinson also flew a sack of mail across the Mississippi River, over to Illinois, and then back to his starting point at the riverbank at the foot of North Market Street, which marked the first interstate airmail service to and from St. Louis.

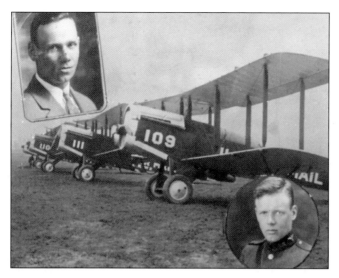

In May 1925, Phil Love (upper left) and Charles Lindbergh (lower right) graduated from the Army Flying School in Brooks, Texas. They were both asked to join the Robertson Aircraft Company at Lambert Field in St. Louis as airmail pilots on routes that had been granted to Robertson by the U.S. government. Lindbergh was hired as the chief pilot and Phil Love as the deputy chief pilot. The Robertson fleet of DeHavilland DH4 mail delivery aircraft is depicted in this photograph.

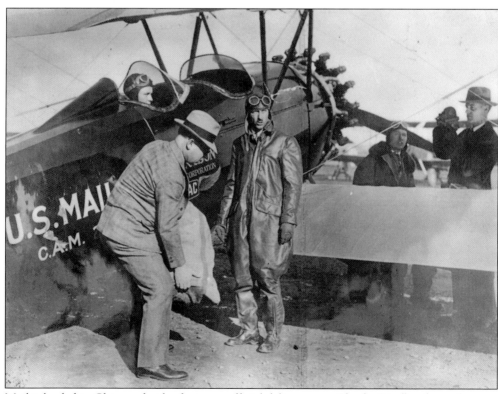

Mail is loaded in Chicago for the first-ever official delivery run. Charles Lindbergh can be seen in the cockpit of one of the Robertson-owned Pitcairn Sport Mailwing aircraft.

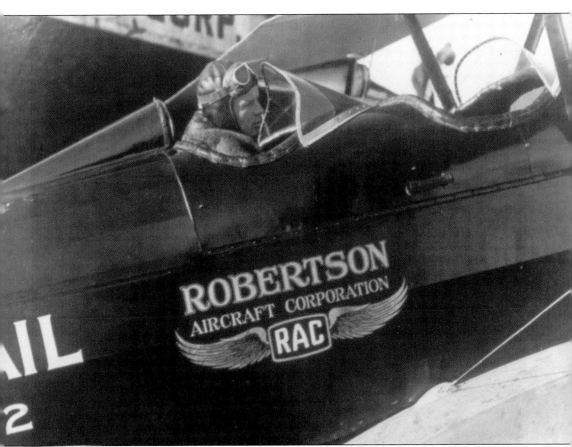

On April 5, 1926, Charles Lindbergh made the first-ever official mail flight flying for Robertson Aircraft from Chicago back to Lambert Field in St. Louis. Robertson won one of the first of the five airmail routes granted to private operators by the U.S. Postal Service. Robertson ran airmail between St. Louis and Chicago, with stops in Springfield and Peoria, Illinois, while utilizing DeHavilland DH4 and Pitcairn Mailwing aircraft. Officially this delivery service was designated Route No. C.A.M. 2 (Contract Air Mail). The airline that William and Frank Robertson had started as an airmail service under contract for the U.S. government, which later offered airline service from Lambert Field under the name of Robertson Airlines flying St. Louis–built Ryan Brougham aircraft between St. Louis and New Orleans, Louisiana, was purchased by Universal Aviation and renamed Universal Air Lines, Inc. The present-day American Airlines Corporation, based south of the Dallas-Fort Worth International Airport, would not have existed if it were not for Robertson Airlines and Universal Air lines, because the sale of these two corporations formed the basis of American Airlines. The scheduled service between St. Louis and Chicago that originally began as an airmail run is arguably the oldest scheduled and continuously running airline route in the United States.

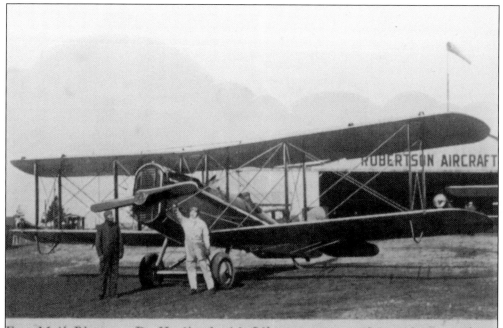

Fast Mail Plane—a De Haviland with Liberty motor, used in Advanced Course

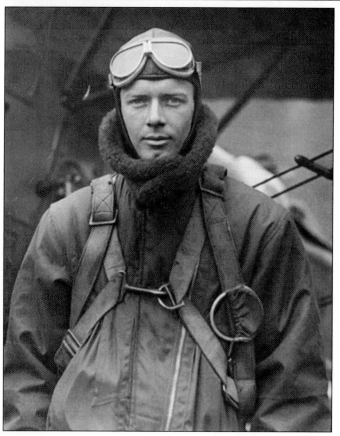

Here is a Robertson DeHavilland DH-4 mail plane. The DeHavilland was designed in Great Britain, but for the U.S. war effort during World War I, various contractors to the U.S. government built it under license. These aircraft were a logical choice for carrying the mail, mainly because of their size and carrying capacity. The U.S. version of the DH4 was powered by the 400-horsepower V-12 Liberty engine produced by Packard in Detroit.

Mail pilot Charles Lindbergh is on duty and ready to fly his night sortie to Chicago.

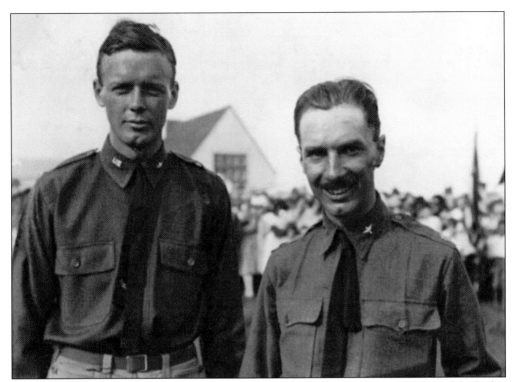

Charles Lindbergh (left) was encouraged to enlist with the 110th Observation Squadron by his friend Maj. Ray Wassall (right). Lindbergh was immediately promoted to first lieutenant upon joining the squadron, which would eventually become the Missouri Air National Guard unit that held a historic and long tenure at St. Louis Lambert International Airport.

Charles Lindbergh hired Bud Gurney (right) to be his chief mechanic for the airmail fleet of aircraft at Robertson Aircraft. Les Smith (left) was one of the airmail pilots working under Lindbergh at Robertson.

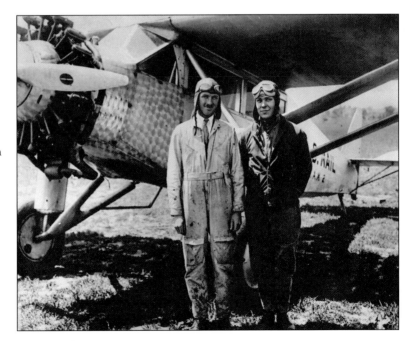

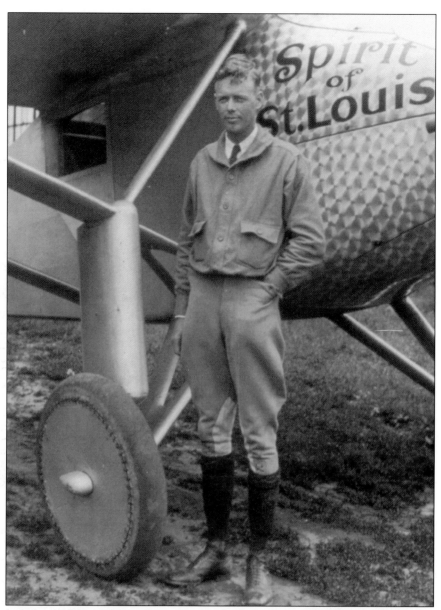

With the help of Albert Bond Lambert, Charles Lindbergh garnered financial backing for his most famous flight from these St. Louis businessmen: Harold M. Bixby, the head of the St. Louis Chamber of Commerce; retired stockbroker Harry F. Knight; Knight's broker son Harry H. Knight, treasurer of the Lambert Pharmacal Company (Listerine); Albert Bond Lambert's brother J. D. Wooster Lambert; *St. Louis Globe-Democrat* newspaper publisher E. Lansing Ray; aviation brothers Frank H. Robertson and William B. Robertson; and insurance broker Earl C. Thompson. In competition for the $25,000 Orteig Prize, Lindbergh would cross the Atlantic non-stop. New York hotelier Raymond Orteig was a visionary who offered a challenge to the world—a non-stop transatlantic flight between New York and Paris. A total of 16 men lost their lives in pursuit of the prize; Lindbergh succeeded. From May 20 to May 21, 1927, Lindbergh flew a solo non-stop trip from New York to Paris—a distance of 3,610 miles—in 33 hours and 30 minutes in the *Spirit of St. Louis.* He returned to the United States along with the *Spirit of St. Louis* aboard the USS *Memphis.*

Lindbergh received the biggest ticker-tape parade ever held in New York City. He also received a ticker-tape parade in St. Louis on June 18 before embarking on a grand tour, jointly sponsored by the Guggenheim Fund and the U.S. Department of Commerce. The tour's purpose was to stimulate public and commercial interest in aviation and its safety and practicality of professionally flown aircraft.

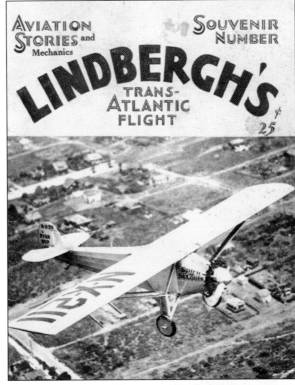

This picture was taken of Charles Lindbergh on the ramp of the 110th Observation Squadron at Lambert Field prior to him embarking on his tour in his *Spirit of St. Louis*. It is believed that 30 million people saw Lindbergh and his aircraft during this tour with stops in 82 cities during a 22,350-mile route that took 260 flight hours to complete.

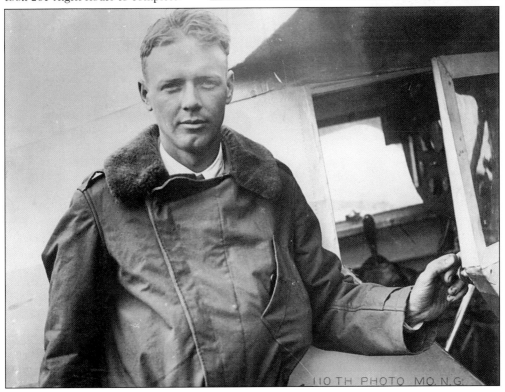

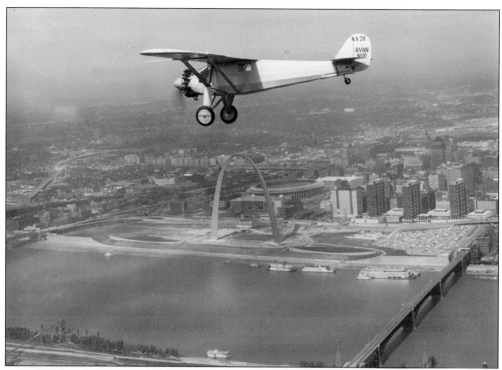

One of the three Lindbergh Ryans made especially for the 1957 motion picture production of *The Spirit of St. Louis* is flown past the recently completed Gateway Arch Monument and Grounds in 1965.

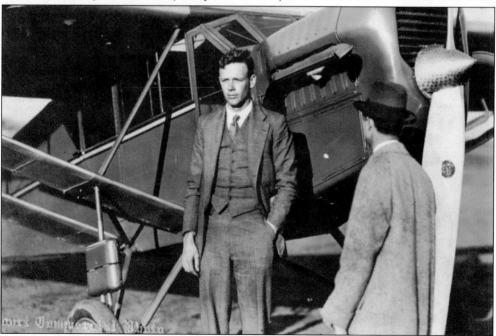

Even though a return to a normal life would be impossible, Charles Lindbergh did stay with Robertson Aircraft, which would soon become Curtiss-Robertson. Here is Lindbergh posing in front of a Curtiss Robin aircraft that he was both testing and promoting on behalf of the company.

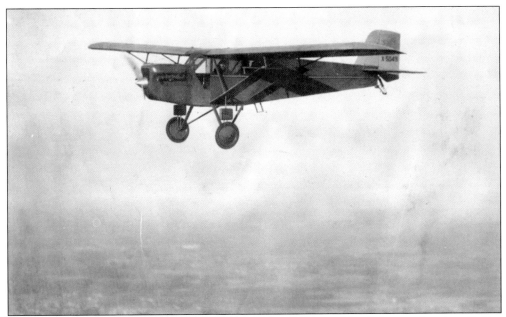

Lindbergh is pictured in flight during certification flight-testing of the *Robin* over Lambert Field.

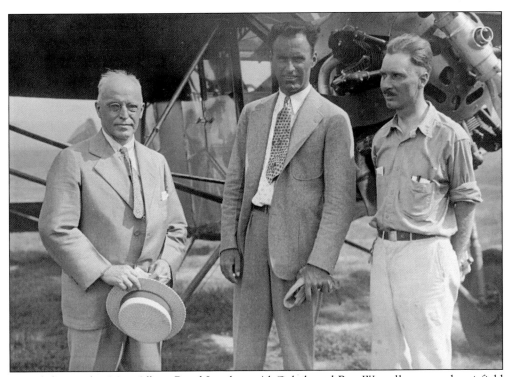

From left to right, Maj. Albert Bond Lambert, Al Gobel, and Ray Wassall pose at the airfield that was founded by Lambert in 1920.

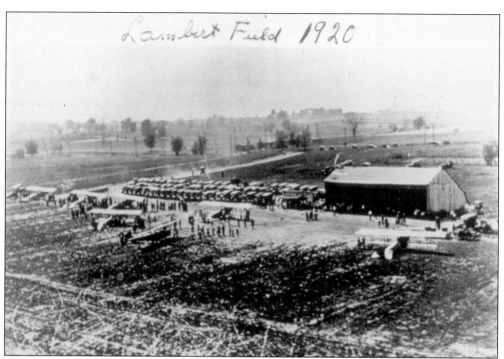

Lambert Field 1920

Major Lambert and his Missouri Aeronautical Society rented this 160-acre cornfield and established an airfield here.

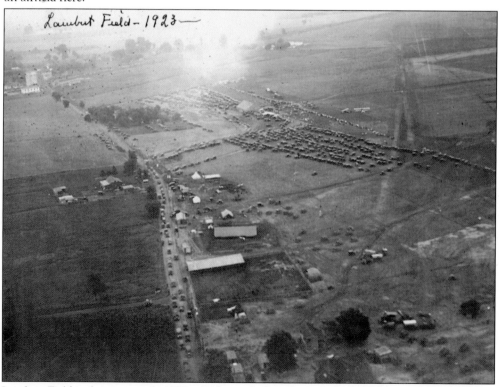

Lambert Field - 1923 -

Lambert Field is shown in early 1923, which was before the International Races were held there.

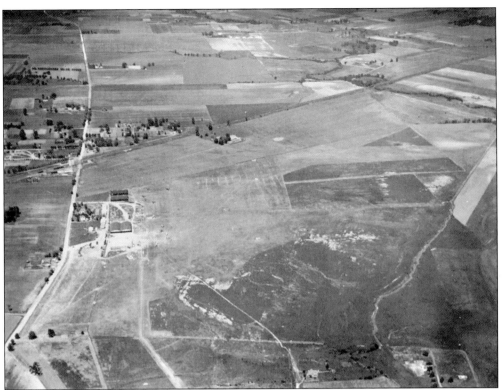

This view shows the National Guard of Missouri Hangers (the dark buildings) and the Robertson Aircraft Hangars at Lambert Field in early 1924.

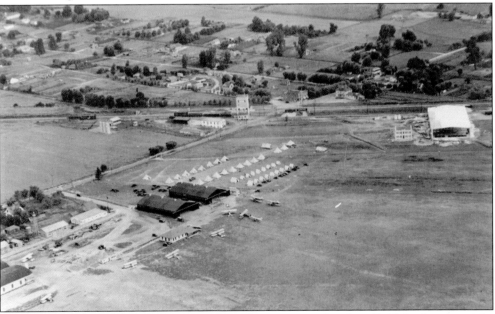

The National Guard of Missouri holds its annual maneuvers at Lambert Field. In the upper-right corner, one can see the U.S. Naval Air Station Hanger under construction. After Lindbergh made his historic flight from New York to Paris, growth and development exploded at Lambert Field.

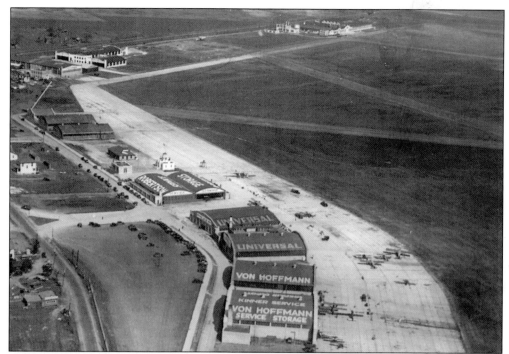

In this photograph, taken in 1929 at the west end of the field close to where the modern-day airport road tunnel now emerges, the Curtiss-Wright Aircraft Manufacturing Plant can be seen at the top right. The U.S. Naval Air Station is visible at the top left; below that, on the left, is the Missouri Air Observation Squadron (Air National Guard); below that, the café, the Airport Administration Building, and terminal can be seen. In the middle foreground are Robertson Aircraft Corporation, Universal Aviation, and Von Hoffmann Aircraft.

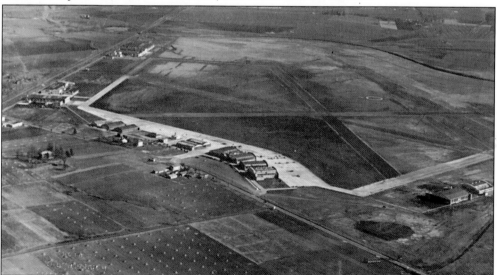

This view looks toward the east over Lambert Field, taken from altitude. By 1929, the City of St. Louis had taken over the ownership and administration of the airfield, at the behest of Albert Bond Lambert, after an agreement was reached for this transfer to occur. The agreement was signed on February 7, 1928.

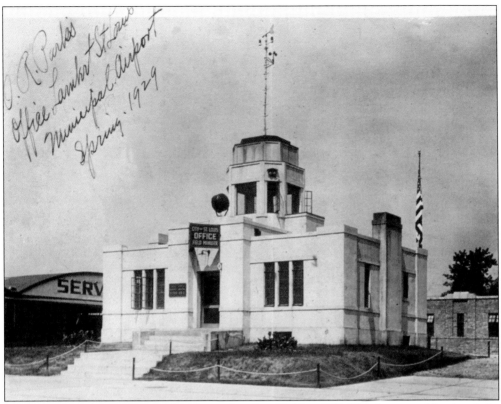

The city built its administration building at Lambert Field in 1928. The directional landing light can be seen on the parapet in front of the control room. By 1929, air traffic in and out of Lambert Field had increased to dangerous proportions without any form of control.

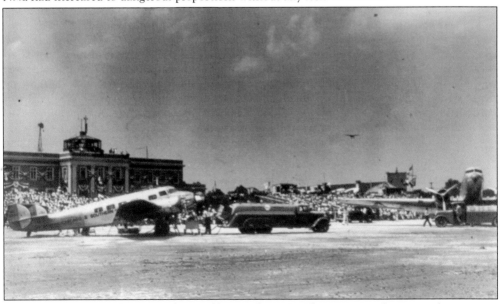

The new terminal building at Lambert Field opened in 1933. The new control room can be seen atop of the building. This room was run by Archie League, the world's first air-traffic controller.

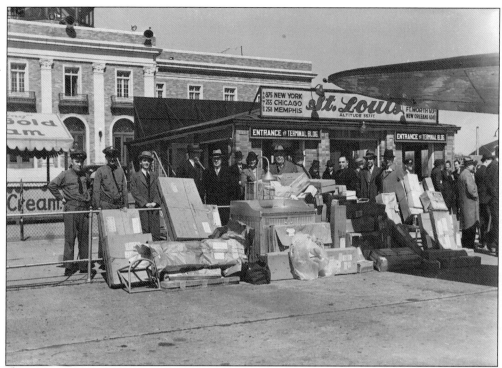

Both cargo and passengers await their aircraft in front of the terminal building in the mid-1930s. The wing tip that is shadowing some of the packages ready for shipment is from a TWA Douglas DC-2.

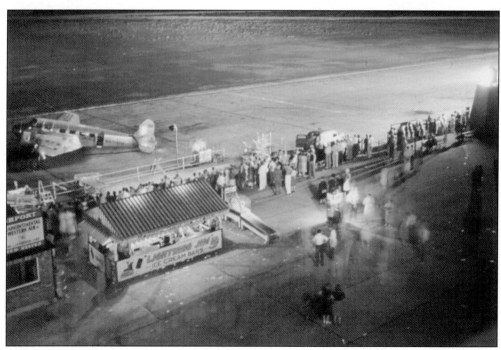

Before everything changed during World War II, going to Lambert Field with a date or one's family on a Friday or Saturday night was very popular.

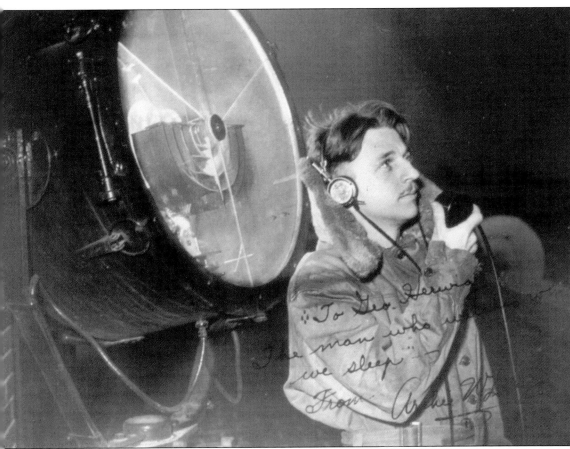

The City of St. Louis appointed St. Louis native Archie League as the air-traffic controller at Lambert Field. This move by the city may have made Archie the world's first air-traffic controller. When he started his job, the tools of the trade consisted of a wheelbarrow, two flags (one red for "hold" and one checkered for "go"), a notepad, pencil, and a stool to sit on. He later received a modern control tower from which to direct traffic in 1933, when the new terminal opened. This image shows Archie on duty at night, speaking to an aircraft that is inbound to the airfield. The spotlight behind him is a landing light that could illuminate the runway for night landings. Much later, in the mid-1960s, Archie became the director of all air-traffic control for the entire nation at his post at the Federal Aviation Administration. The FAA gives an annual award in his name to air-traffic controllers who demonstrate acts of supreme aviation safety.

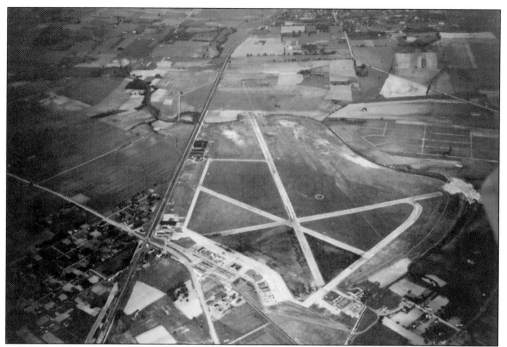

Above the white excavation area, to middle right of this photograph, is the approximate present-day location of the current airport terminal and ramp. On July 12, 1930, Adm. Richard E. Byrd formally dedicated the Lambert-St. Louis Municipal Airport.

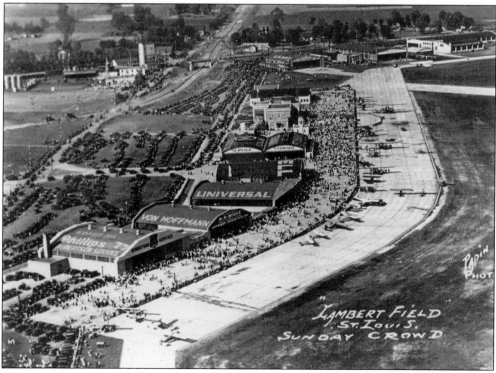

This usual Sunday crowd of onlookers is shown in a view taken from the air in 1935.

Three

THE AIRPORTS
OF ILLINOIS

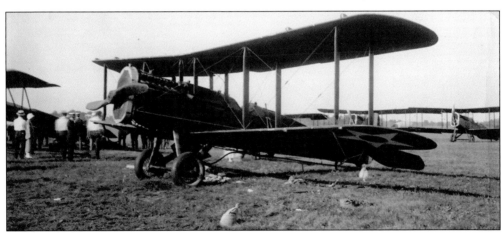

On September 1, 1917, Scott Field in Shiloh, Illinois, officially opened under a congressional decree as a mammoth Army Air Corps Aero Squadron flying school. The field was named after Cpl. Frank S. Scott, who was the first enlisted man in the United States to lose his life in an air accident. Here Liberty engine-powered Jennys await their instructors and cadets to take them into the air.

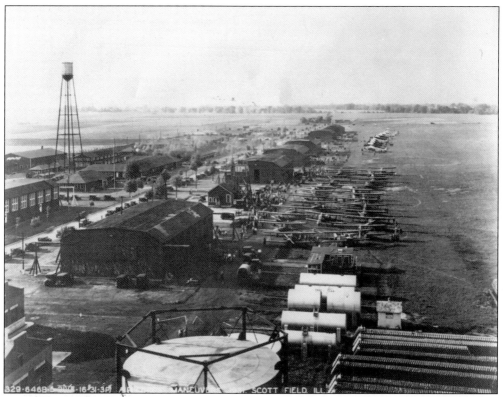

This photograph is of Army Air Corps maneuvers at Scott Field. The world's second-largest building ever constructed at that time, the dirigible and airship hangar at Scott Field was completed in 1922 with a total cost estimated at $1,360,000, which was a staggering amount for the time.

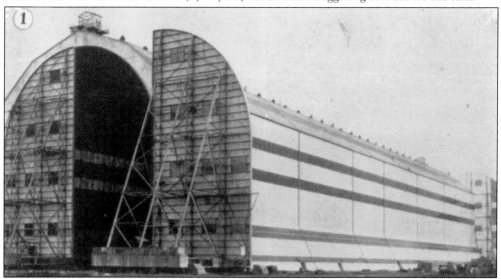

It took 351 days to build the 11.3-acre (20,000-cubic-foot) building; one million rivets held together the 195-railcar loads of steel that weighed 4,050 tons. Additionally, 290 railcar loads of cement, 160 railcar loads of cinders, 25,000 ceramic tiles, 27,000 square feet of glass, and 134,000 square feet of siding material was used to construct this magnificent building. Each of the four doors weighed 650 tons.

Here is the airship mooring mast at Scott Field. Each airship was equipped with a clip-ring mooring device on the nose (front end) of the ship. This was then clipped and tied to the swiveling mooring post at the top of the tower. If one looks closely, a mooring technician standing at the top of this tower can be seen ready to secure a docking airship.

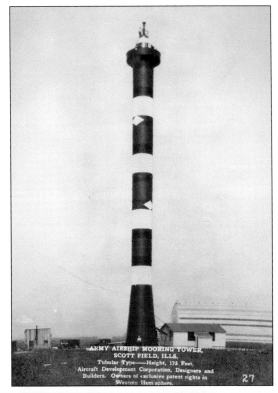

A C Class semi-rigid Army Air Corps dirigible is seen emerging from the Scott Hangar. The mooring cap is at the front of the aircraft. Just aft of the nose, one can see the pressure-relief valve that would relieve gas pressure as the airship climbed, to both maintain rigidity and structural integrity. The control-cockpit gondola is obscured in this picture by the ground crew, who are walking this airship back into the hanger.

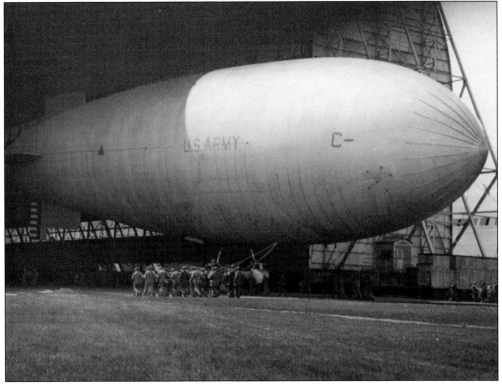

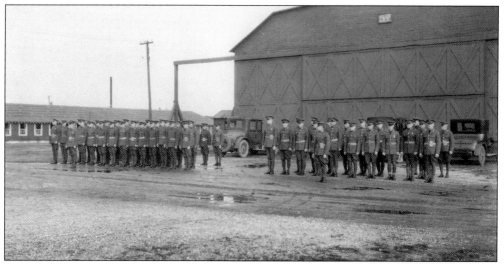

Here is the 24th Airship Service Company at Scott Airfield in 1927. These men were adjunct to three other airship companies—the 8th, 9th, and 12th—that all came under the group command of the 21st Airship Group headquartered at Scott Field. The Air Service of the U.S. Army at this time had about 1,000 officers commanding almost 10,000 enlisted men. In addition to their particular service rank, the officers were either rated as airplane pilot, airplane observer, airship pilot, airship observer, or balloon observer.

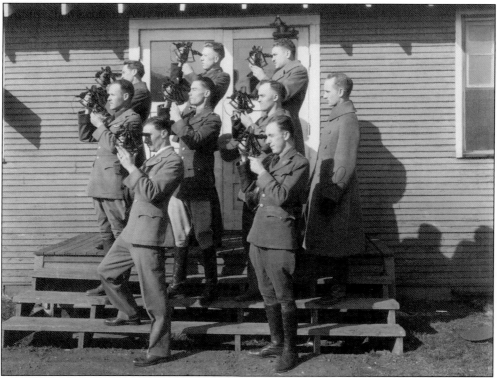

Navigator cadets at Scott Field use sextants to determine the sun's azimuth so they can compute their exact position on a navigational chart. Celestial navigation has pretty much been replaced by the Global Positioning System provided by geostationary satellites in space.

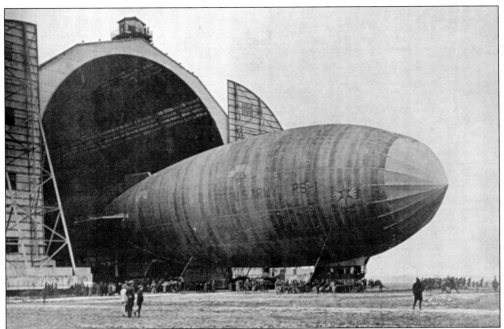

An RS-1 semi-rigid Army Air Corps dirigible is emerging from the Scott Hangar. The RS-1 is the largest airship model that was used at Scott Field. The backbone structure of this vessel consisted of an internal "A" frame that held fuel and water-ballast tanks along its beam length. The control, or cockpit gondola, is seen at the front of the craft, while the engine pod was at the rear portion of this behemoth. The engine pod housed 400-horsepower Liberty V-12 engines that drove reversible propellers.

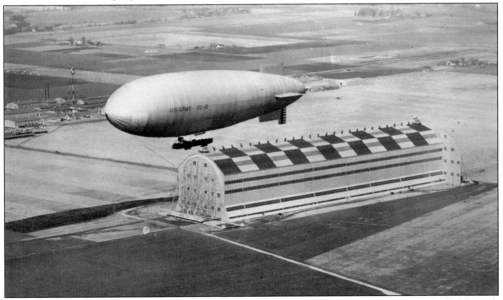

An XTC-6 airship is seen over the Scott Hangar. The airship cadet program consisted of a 12-week program of air and ground training. A typical airship crew consisted of a pilot, co-pilot, and two engineers. Obviously there was a small army of ground crew necessary to be on hand for takeoff, landing, mooring, and ground maneuvering.

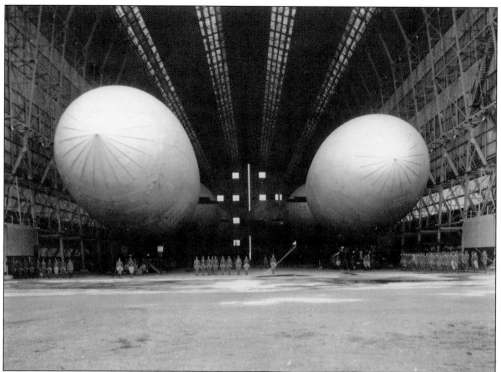

This view was taken inside the massive airship and balloon hangar at Scott Field.

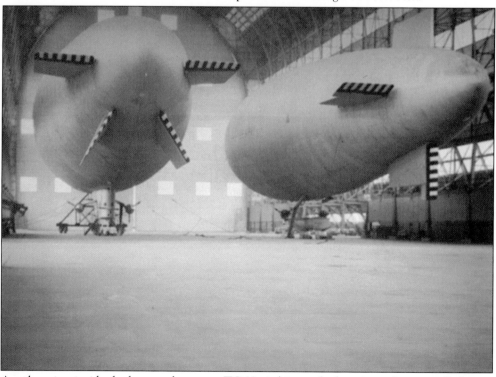

Another view inside the hangar shows two TC-10 airships parked awaiting their next flight.

54

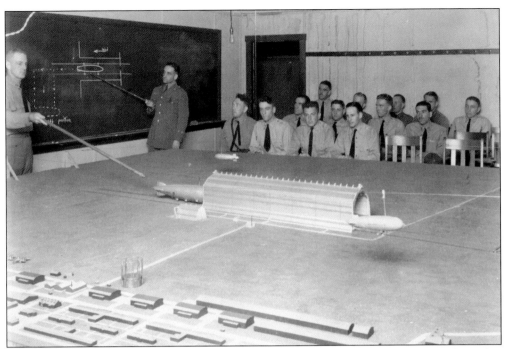

An airship cadet class is in session in one of the training rooms at Scott Field. Here the cadets are being shown the landing, ground maneuvering, and parking procedures for returning the airship to the hanger.

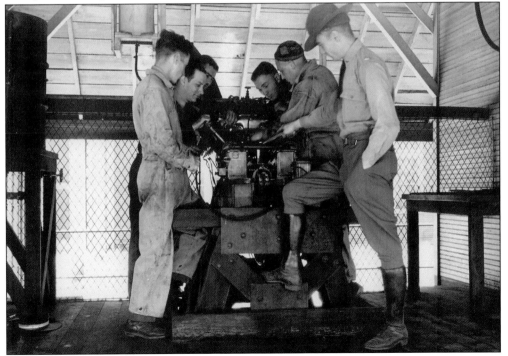

Army Air Corps students work on one of the Liberty airship engines at one of the engine testing stands.

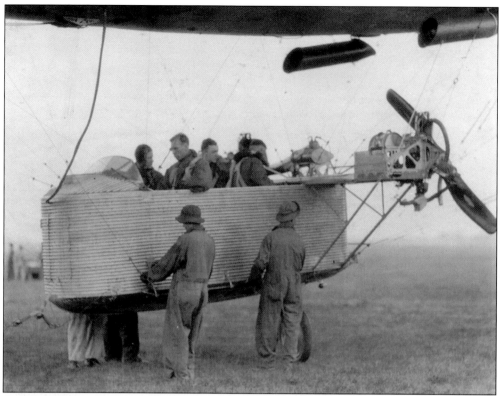

A Scott Field ground crew manhandles one of the school's TC-10 dirigible semi-rigid airships toward the hangar. The two pipes hanging beneath the gasbag are ram air spouts that could be lowered into the airstream as the craft flew forward. This enabled the craft to retain and maintain suitable gas pressure (there were separate gas and air bags contained within the airship balloon structure).

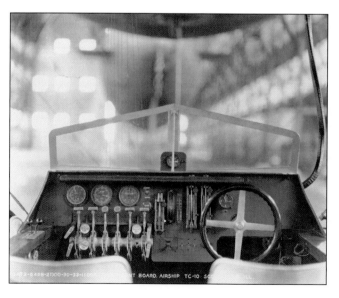

This craft had two engines; one was located on the left side of the cockpit in the gondola. This photograph is of one of the Scott airship school's TC-10 airships, with the duel throttle, mixture, and propeller seen. Above these are the airspeed and revolutions-per-minute gauges and the altimeter. In the middle are gas pressure gauges, a clock, a thermometer, and spirit level fuel gauges. The tiller and wheel (control column), which are similar to a car's system, control the tail surfaces that steer the craft.

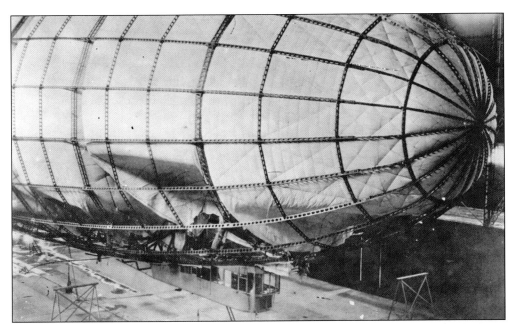

An airship undergoes repairs in the hangar. The closeness and the number of thread stitches per inch vary depending on the gas pressure and structural integrity required in different areas of the gas envelope. The closer that they are means that higher pressures can be sustained. The exterior skin of this airship has been removed for this maintenance operation. The aluminum girder structure acts like a skeleton, providing rigidity and strength.

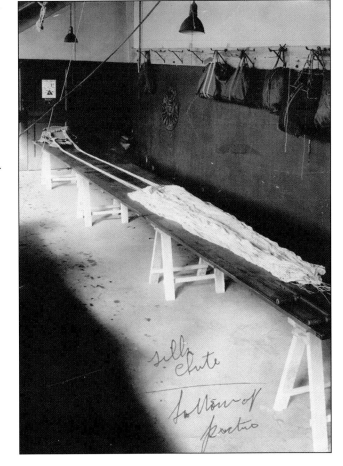

A parachute undergoes inspection at Scott Field. Once it has passed muster, the parachute canopy and tether lines shall be carefully folded and repacked into a shoulder-worn parachute bag.

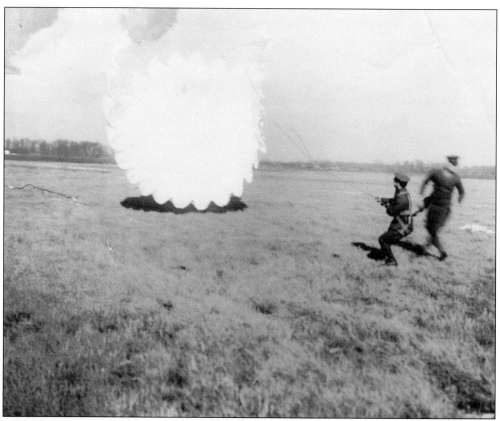

Here a parachute is being tested as part of its inspection prior to repacking. A parachute is packed so that it will withdraw from its bag without tangling.

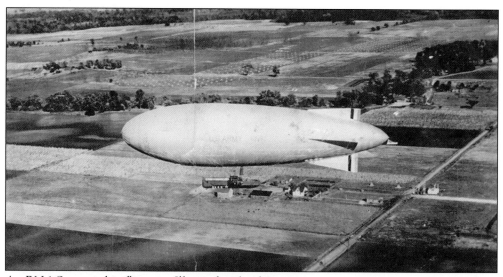

An RN-1 Scott airship flies over Illinois farmland.

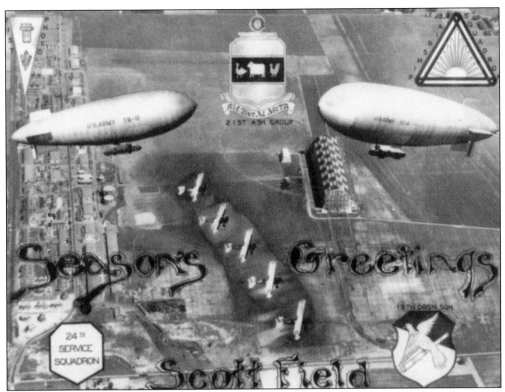

This Scott Field Christmas card dates from December 1931, eight years before the balloon and airship squadrons were deactivated and the massive hanger demolished.

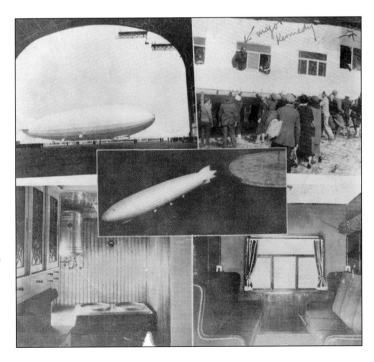

The USS *Los Angeles* arrives at Scott Field. The Zeppelin ZR-3–class airship was built in Friedrichshafen, Germany, in 1924 and became the U.S. Navy's most successful airship, which was used for testing defense delivery systems. She was scrapped in 1937. Her helium gas capacity was 2,472,000 cubic feet. Maj. Frank M. Kennedy (marked by an arrow), later a lieutenant colonel, was the commanding officer of Scott Field from September 1921 to February 1922 and from August 1933 to March 1937.

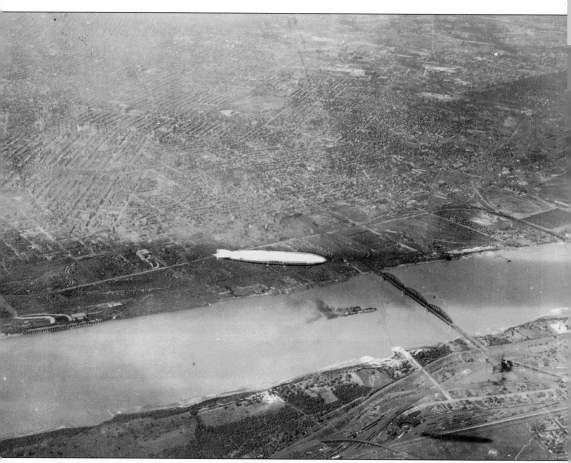

The USN *Shenandoah* flies over McKinley Bridge in 1923. This is an earlier Zeppelin ZR-1 airship that was built in post–World War I Philadelphia after the U.S. Navy procured technical data, parts, and machining processes from the defeated German government. Built in 1923, the *Shenandoah* boasted a helium gas capacity of 2,115,000 cubic feet. The zeppelins were all designed to be buoyed by hydrogen gas instead of helium, but the U.S. government knew the dangers of hydrogen and consequently converted their Zeppelin-class airships to helium.

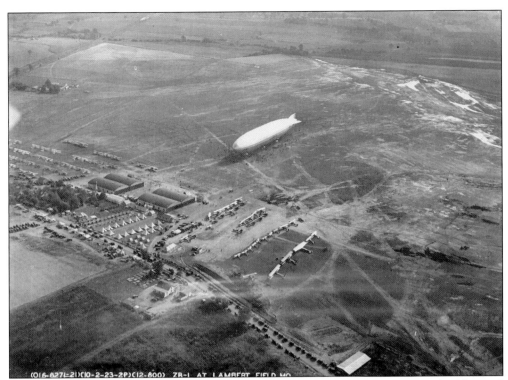

(O16-8271-21)(10-2-23-2P)(12-800)   ZR-I   AT LAMBERT FIELD MO

The *Shenandoah* is pictured at Lambert Field in 1923 during an Army-Missouri Air National Guard exercise. In September 1925, less than two years after this photograph was taken, the USN *Shenandoah* visited Scott Field; she was lost on the following day in violent storms encountered in Ohio on her way back to her base in Lakehurst, New Jersey.

To the east of St. Louis, in Cahokia, Illinois, Oliver Parks established his air college. Parks was a Chevrolet dealer before he decided to open an air-training school, and he would use personal aircraft to promote car sales. The lure of aviation was too great, and after he received his transport pilot certificate, he left the automobile business and invested his money in an aviation business and school.

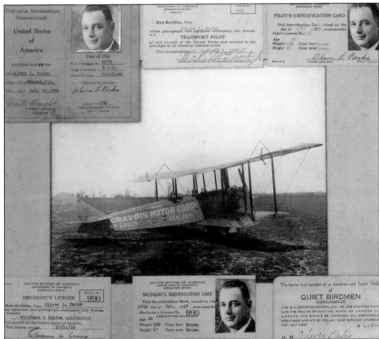

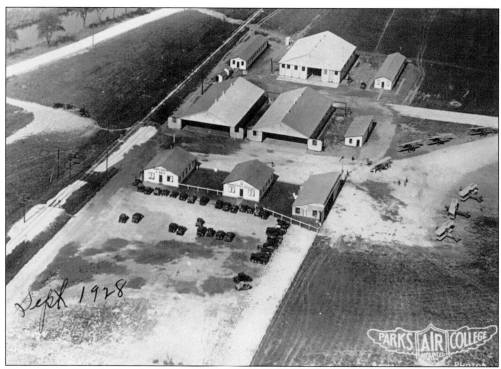

Oliver Parks founded his Parks Air College in East St. Louis/Cahokia, Illinois, in 1927. Parks Air College became the first federally approved school of aeronautics, receiving Air Agency Certificate No. 1.

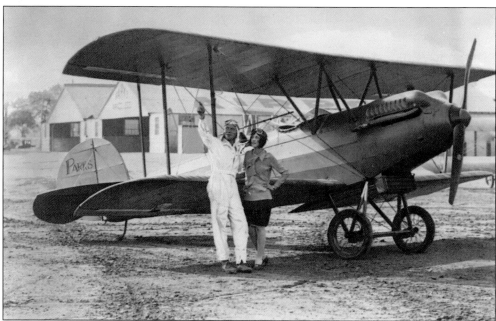

Parks built his own training aircraft for his college in 1929. One of his P-1 trainers is seen posing with a student pilot, who was also his girlfriend, for a photograph that was used in an advertisement for the school that appeared in various periodicals.

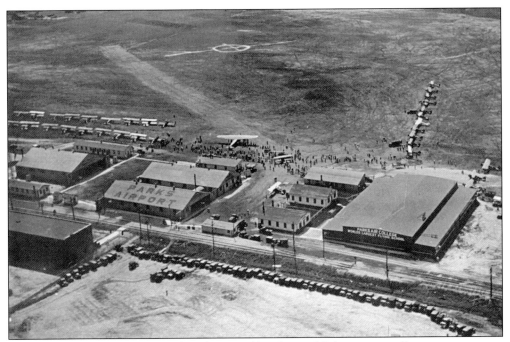

By 1936, Parks Air College was identifying itself as the world's largest air college.

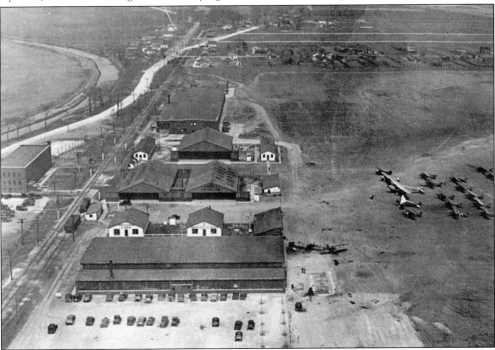

In a 1934 photograph of Parks College, a Curtiss-Steinberg Airport hangar can be seen in the top right-hand corner. In 1997, after the administration of the college was transferred to St. Louis University, the Cahokia campus was closed; its most recent hanger site was converted into a YMCA facility. The school moved into new facilities located on the main St. Louis University campus across the Mississippi in downtown St. Louis.

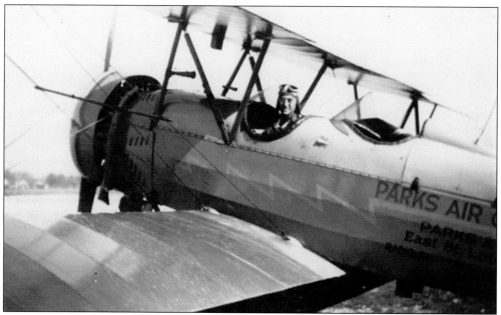

During radio, instrument, and night-flight training, the instructor sat in the front cockpit and acted as the student's mentor and also as the student's lookout for other aircraft. The student pilot sat in the rear cockpit that was covered by an instrument hood that snapped to the inside of the cockpit rim. The hooding rendered the student blind to the outside world and forced him to rely entirely on his instruments.

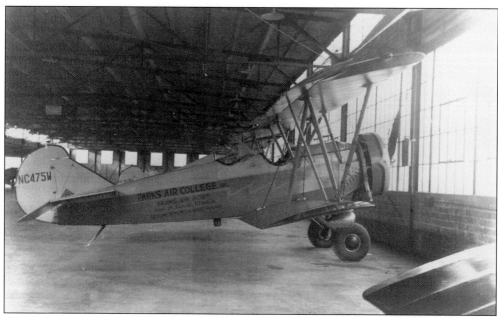

A Travelair trainer is parked in the college hangar for the night.

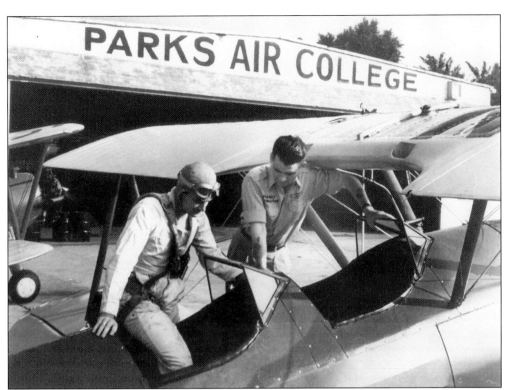

PARKS AIR COLLEGE

In a photograph taken shortly before the United States entered World War II, a student and instructor prepare to go aloft in one of the college's PT-13 aircraft. Flight training is still undertaken on the "east side," which is now at the St. Louis Downtown Airport (formerly Curtiss-Steinberg Airport).

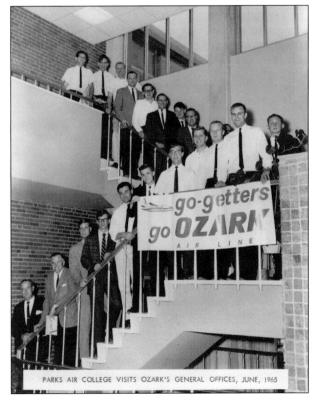

PARKS AIR COLLEGE VISITS OZARK'S GENERAL OFFICES, JUNE, 1965

Parks College students visit the Ozark Airlines' St. Louis offices in 1965. This photograph was taken 19 years after the then soon-to-retire Oliver Parks had given his college to St. Louis University. His college was renamed by the university as the Parks College of Engineering, Aviation, and Technology of St. Louis University.

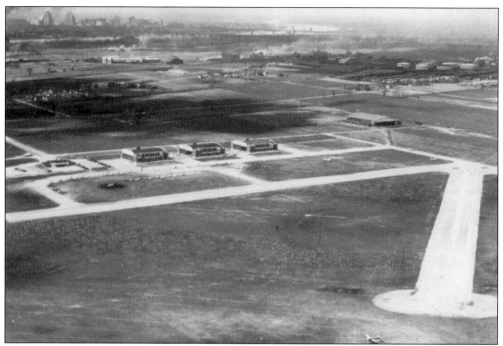

In May 1929, the Steinberg Airport opened in Cahokia/Sauget, a mile from Parks Air College Field. Designed for Transcontinental Transportation Corporation (TAT, the forerunner to TWA) as a headquarters and overhaul base for its Ford Tri-motor fleet, the Curtiss-Wright Corporation purchased the airport before TAT even began operations. The Eads Bridge is visible at the top center. Due to an electrical fault, the third hanger, which is seen on the right, burned to the ground.

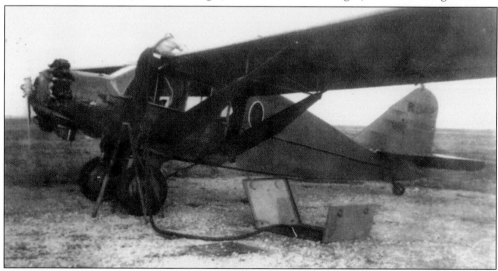

A Bellanca Sky Rocket is being fuelled at Curtiss-Steinberg Airport. During a 1930 air show here, Jimmy Doolittle's Travel Air Mystery Ship that had been newly modified at Parks College broke up during a high-speed pass down the runway. At 300 miles per hour, Doolittle (of Shell Petroleum's St. Louis-based aviation department) experienced total disintegration of his wings. He rolled inverted and bailed out at 400 feet. His parachute deployed and he survived. His aircraft crashed near the perimeter fence.

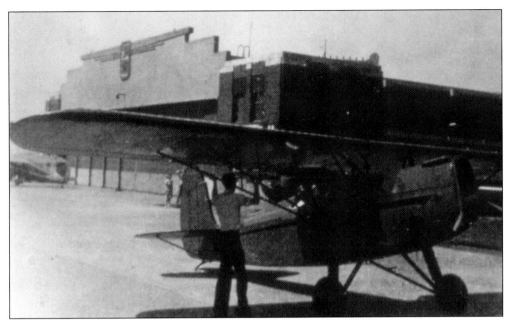

Hangar Two at the Curtiss-Steinberg Airport, pictured in the mid-1930s, is now home to the Greater St. Louis Air and Space Museum. Today called the St. Louis Downtown Airport, in 1940 Oliver Parks purchased the Curtiss-Steinberg Airport and renamed it Parks Airport. In 1959, the airport closed while permission was sought to use the site for housing. It was purchased by the Bi-State Development Agency and renamed Bi-State Parks Airport, reopening in 1965.

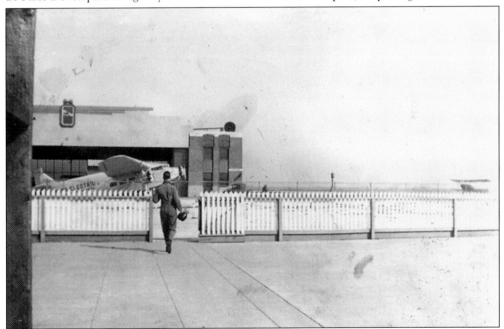

This Ford 4-AT-B Tri-motor survey aircraft was based at Curtiss-Steinberg Airport. Union Electric Corporation purchased this plane in May 1930 for survey work for a hydroelectric dam project; it was later used for corporate transport and was sold in 1937, replaced by a used Boeing 247D. This Ford Tri-motor is on display in the National Naval Aviation Museum at Pensacola Naval Air Station.

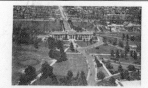

Jefferson Memorial, Home of the Lindbergh Trophies

THE "HOME TOWN" TOUR

To give you a better idea of how your own home looks from the air, this trip is designed. From Curtiss Field over the Mississippi and south along Broadway and the River to Carondelet Park. North over Kingshighway Memorial Boulevard passing Tower Grove Park, Shaw's Garden, and Forest Park, to hover over the old Fairgrounds and Sportsmans Parks. Returning to the field you pass over downtown St. Louis and across the River for a view of East St. Louis. Here is an opportunity to really see the cities of St. Louis and East St. Louis.

FARE PER PASSENGER $5.00

THE "GREATER ST. LOUIS" TOUR

For those to whom industry is appealing and beauty fascinating, this trip will prove most enjoyable. North over East St. Louis and the National Stock Yards to the tri-cities—Madison, Venice and Granite City, with their enormous factories and webs of steel rails. Thence over the Chain of Rocks, the source of the St. Louis water supply, to view the Missouri emptying her muddy waters into the Mississippi. Returning, you see the old towns of Florissant, Ferguson and Normandy and fly over the northwest industrial section, Sportsmans and Old Fairgrounds Parks to the business center of town with its new Courthouse and Memorial Plaza. Thence across the river and back to the field.

FARE PER PASSENGER $10.00

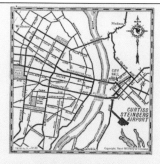

MAP of ST. LOUIS

SHOWING THE ROUTES TO
CURTISS-STEINBERG AIRPORT

Busses and Trolley Cars leave Eads Bridge on Regular Schedule.

By motor across the Municipal Bridge and turn left one block to Highway No. 3, turning left on the highway to Curtiss-Steinberg Airport, or, via the Eads Bridge, east on Broadway to 8th Street in St. Louis and right on highway No. 3.

A Plane for Every Purpose

Air Tours—Charter Flights—Sightseeing Trips—Air Taxi Service—Mapping and Aerial Survey—Day and Night Service.

CURTISS-WRIGHT FLYING SERVICE
"World's Oldest Flying Organization"

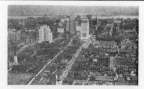

Downtown St. Louis and Mississippi River from the Air

THE "METROPOLITAN TOUR"

This trip makes a delightful short hop over Cahokia Power Plant and the Mississippi River to downtown St. Louis. Here you get a close up of our new Courthouse, Memorial Plaza and towering office buildings from a new angle. You are thrilled by the appearance of widened Olive Street and other thoroughfares stretching out like ribbons to the distant residential sections of our city.

A most enjoyable and inexpensive ride that will give you a new pride in your city.

FARE PER PASSENGER $3.00

THE MISSISSIPPI MERAMEC VALLEY TOUR

For a cruise combining many interesting points, with the grandeur and beauty of the mighty Father of Waters and the cool inviting Meramec Valley, this trip is ideal. South over the old village of Cahokia, across Arsenal Island within sight of the Anheuser-Busch plant and on down the Mississippi bluffs past the Jefferson Barracks, U. S. Veterans Hospital and Koch Hospital, to the mouth of the Meramec. Then up the valley of this meandering stream, flying over Valley Park. Turning north over numerous country clubs and estates, you approach Clayton and come east over Washington University, Forest Park and St. Louis University. You are now hovering over downtown St. Louis and, crossing into Illinois again, you see East St. Louis and return to the field to complete a most interesting aerial tour that will be remembered always.

FARE PER PASSENGER $15.00

A brochure commissioned by the Curtiss-Wright Company publicizes its airport and flight operations. The Curtiss-Wright Corporation purchased the airport from Mark Steinberg and renamed it the Curtiss-Steinberg Airport, pictured here in 1935.

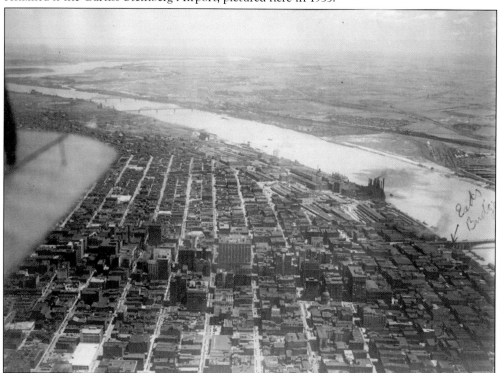

A pleasure flight from the Curtiss-Steinberg Airport cruises over downtown St. Louis. Eads Bridge is indicated by an arrow. The Railway Exchange/Famous-Barr Building is the prominent structure visible in the center of this photograph.

Four

THE AIRLINES
OF ST. LOUIS

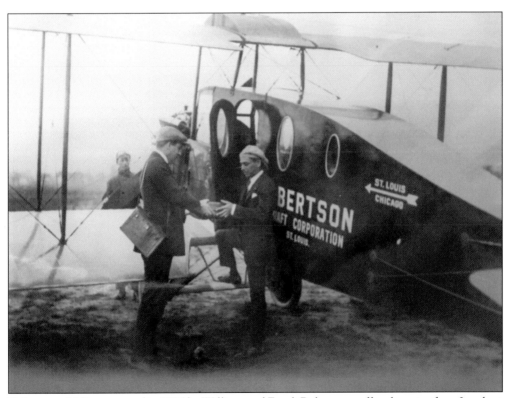

Robertson Airlines, started in 1926 by William and Frank Robertson, offered service from Lambert to New Orleans via St. Louis–built Ryan Broughams. Robertson was purchased by Universal Aviation and renamed Universal Air Lines, Inc. The sale of the Robertson and Universal corporations formed the basis of American Airlines. The scheduled service between St. Louis and Chicago, started as an airmail run, is the oldest, continuously running, scheduled airline route in the United States.

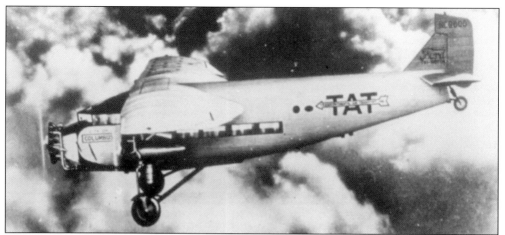

On July 8, 1929, Transcontinental Air Transport offered 48-hour, coast-to-coast service. A train departed Pennsylvania Station in New York that travelled to Columbus, Ohio, where passengers boarded a Ford Tri-motor that stopped in Indianapolis, St. Louis, Kansas City, Wichita, and Waynoka, Oklahoma. Passengers then boarded a train for Clovis, New Mexico, and from there a flight stopped in Albuquerque, Winslow, and Kingman, Arizona, arriving at Grand Central Terminal in Glendale, California, in the late afternoon of the second day after departure.

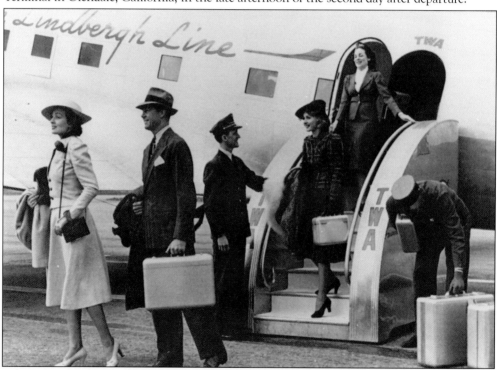

In 1930, Transcontinental Air Transport (TAT), Maddux Airline, and Western Air Express merged, forming Transcontinental and Western Air, Inc. (TWA). In 1945, TWA received its certificate from the U.S. Civil Aeronautics Board, becoming a certified international air carrier, changing the corporate name to Trans World Airlines (TWA). TAT paid Charles A. Lindbergh to recommend potentially profitable airport destinations nationwide, establishing the route structure—the Lindbergh Line. Passengers are seen disembarking a TWA Douglas DC-2 at Lambert Field.

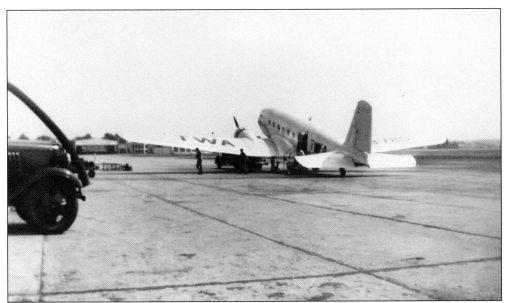

A TWA Douglas DC-2 sits on the ramp in St. Louis. The DC-2 had a crew of three and carried 14 passengers. With an un-pressurized cabin, it normally cruised under 10,000 feet at a speed of about 170 miles per hour. Later TWA moved to the DC-3; both engines on the plane were boosted from 855 to 1,000 horsepower. The final production version of the DC-3 had a longer and wider fuselage than the DC-2, wider wingspan, and stronger undercarriage, carrying 21 to 28 passengers plus additional freight or baggage.

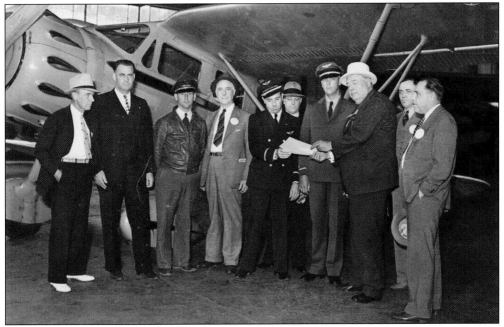

St. Louis Airlines formed and began offering an "on-demand" charter service starting in 1939. Joe Imeson is seen in his uniform in the center of this photograph accepting the Civil Aeronautics Authority Common Carrier Certificate from the then-mayor of St. Louis, Bernard F. Dickmann (in the white fedora hat).

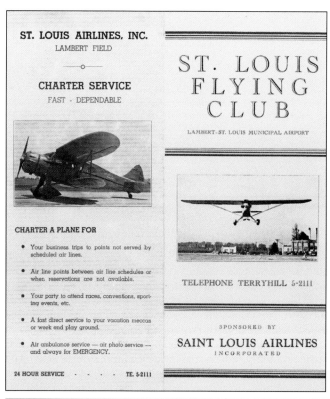

This brochure promotes both St. Louis Airlines and the St. Louis Flying Club.

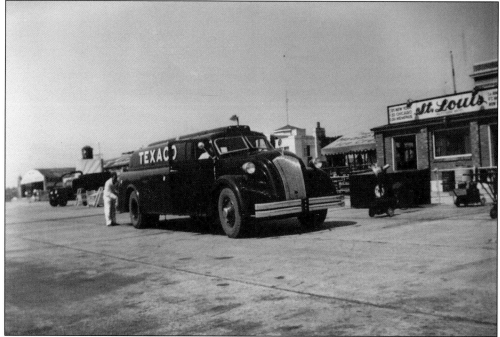

A TWA fuel truck is on the ramp of what had been renamed the St. Louis-Lambert Municipal Airport. The administration building can be seen in the background above the radiator grill of the fuel truck.

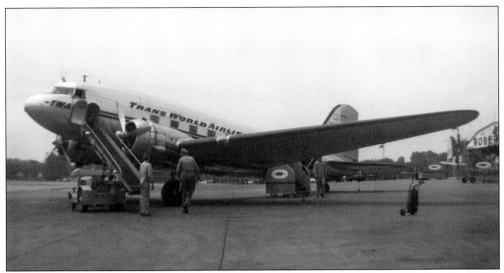

A TWA DC-3 awaits passengers at St. Louis-Lambert Municipal Airport. The Douglas DC-3, developed in 1935, was basically a stretched DC-2, and both American Airlines and United Airlines jumped on this new design. TWA and Eastern Airlines both followed closely behind in their purchases of many DC-3 aircraft, and the first TWA DC-3 entered service with the airline in 1936. To correct a tendency to enter a spin when flown on one engine, engineers at Douglas made a bigger vertical fin, enabling the DC-3 to rudder out of a spin. They also re-designated the single-engine climb speed in the flight manual to prevent stalling if the aircraft was flown correctly.

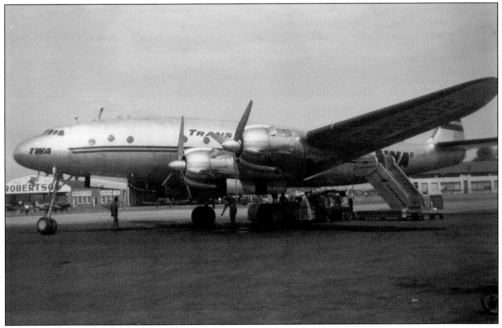

A TWA Constellation sits patiently on the ramp in St. Louis. Each of its four Wright R-3350 Turbo Compound, 18-cylinder, supercharged, radial engines could produce 3,250 horsepower. The most powerful version of the Constellation could carry a useful load of 65,300 pounds, while it could be loaded up to a maximum takeoff weight of 137, 500 pounds.

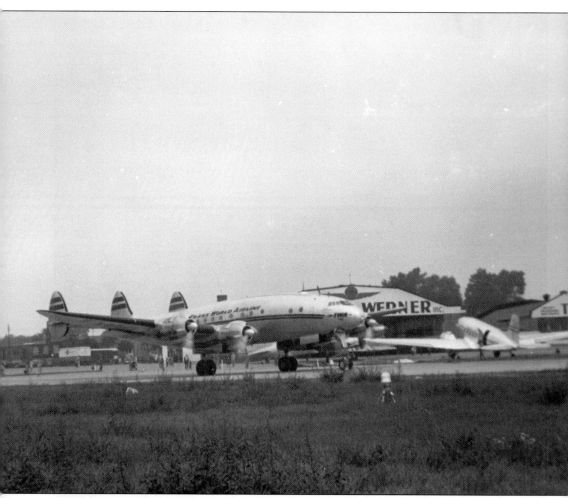

A TWA Constellation taxis for takeoff from Lambert in 1948. The Constellation was designed and built by the Lockheed Corporation. The first production model rolled off the line in Burbank, California, in 1944. This revolutionary design was engineered with the oversight and input of then president and chairman of TWA Jack Frye, while famed engineer/businessman/test pilot Howard Hughes assisted him. Both conceived the idea that for TWA to beat its competitors, it needed revolutionary transport aircraft that could fly coast to coast non-stop in eight and a half hours. The initial design was drawn up for the U.S. Air Force, but with Frye's and Hughes's input, the aircraft soon was re-engineered to become a pressurized airliner that could carry 40 passengers in great comfort over a route of 3,500 statute miles. Its cruise speed was awesome in its day, as it could comfortably fly at more than 300 miles per hour and up to 24,000 feet. The coming of the "jet age" is the only reason that the Constellation became obsolete.

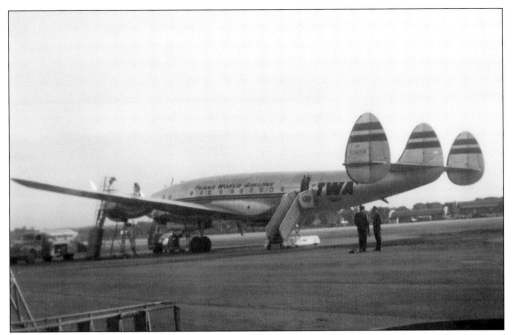

The co-pilot and flight engineer of this Constellation chat while airline mechanics check one of the engines; the aircraft is between passenger flights at Lambert Airport. Needing to carry more passengers and freight on short connecting routes, TWA shopped for aircraft that could fly for less than the Constellation but were larger and faster than its DC-3 and DC-4 fleet.

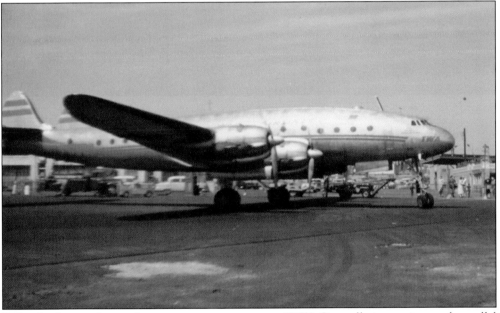

Before loading up for another cross-country segment, a TWA Constellation taxis to park parallel with the St. Louis terminal gate to offload passengers, luggage, and freight. To fly for less money than the Constellation, TWA placed the first Martin 202 into service in 1950; the aircraft carried 36 passengers in an un-pressurized cabin at 220 miles per hour and was powered by two Pratt and Whitney Wasp engines, each producing 2,400 watts for short durations.

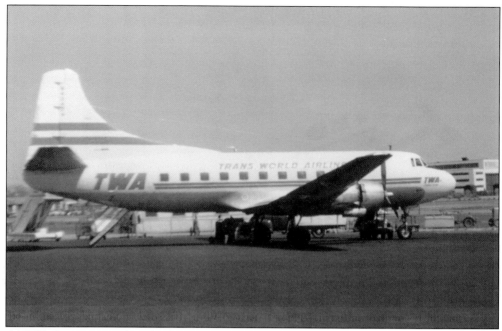

A TWA Convair is ready to load up on the ramp. By 1951, TWA had a 32,000-statute-mile route structure across four continents, and all of this was established before any jet service was available.

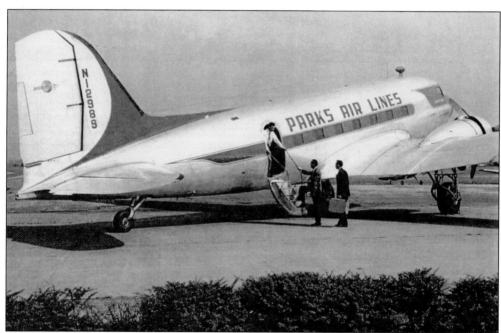

In 1950, would-be carrier Parks Airlines, Inc. of East St. Louis failed to organize to meet the deadline set by the Civil Aeronautics Board. This failure to meet requirements became the foundation for Ozark Airlines. The Ozark Company was incorporated in 1943, when two bus operators in Springfield, Missouri, merged. The company became an airline on January 10, 1945, when it commenced service with Cessna T-50 and Beechcraft F-17D Staggerwing aircraft.

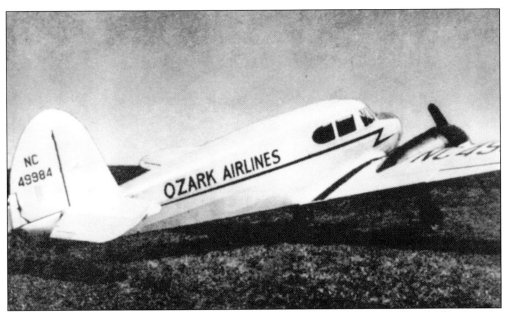

The airline was "brought back to earth" in 1946, when it stopped flying due to financial difficulty. The Cessna T-50 Bobcat, known as the "Bamboo Bomber," was built in Wichita and designed to serve as a military trainer to transition single-engine combat pilots over to flying twin-engine combat aircraft like the P-38 Lightning. It had a range of 750 statute miles and could carry five passengers.

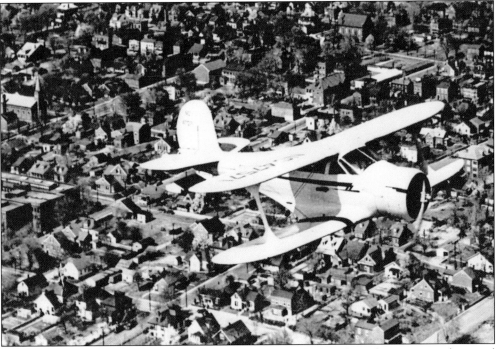

Here is an Ozark Beechcraft Staggerwing in 1945, shortly before the airline temporarily ceased operations. The Beechcraft Staggerwing could only fly three passengers and had a maximum range of 670 statute miles. The choice of using small aircraft with low capacities contributed greatly to the failure of the first Ozark start-up.

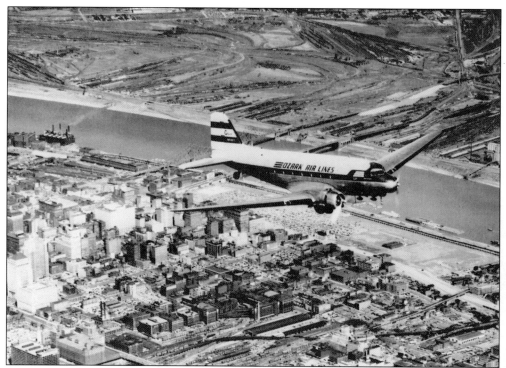

After Parks Airlines failed to receive its operating certificate, Ozark Airlines reorganized and purchased the assets of the start-up corporation. In 1950, it restarted airline service from St. Louis–Lambert Municipal Airport. The Douglas DC-3 was the right size and capacity for this fledgling airline to quickly find success. The DC-3 also provided the first in-flight sleeping berths; Douglas marketed the DC-3 as the "Douglas Sleeper Transport."

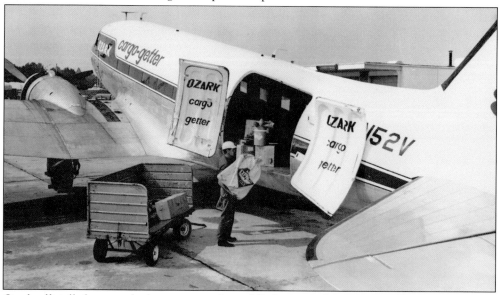

Ozark officially became the hometown airline of St. Louis, with corporate offices established at Lambert Field. Ozark also offered cargo services in some of its Douglas DC-3 fleet. These aircraft had a 6,000-pound cargo capacity, making them very competitive on price.

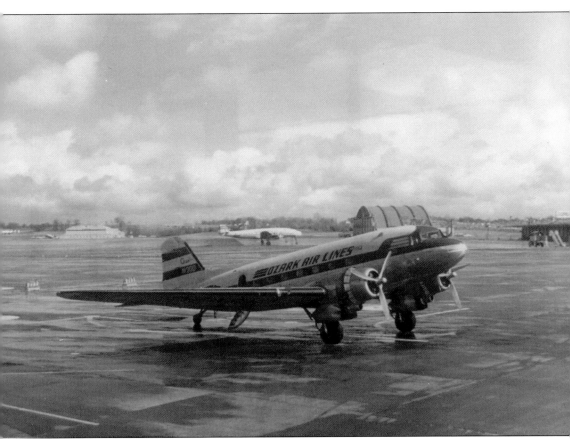

An Ozark DC-3 sits on a wet ramp in St. Louis. This DC-3 is obviously an ex-American Airlines; one can tell by identifying the side on which the passenger door is located. When working with Douglas on the design upgrade of the DC-2, C. R. Smith, the president of American Airlines, insisted that a right side door be installed on the DC-3 instead of the traditional left-side placement. Smith's rationale was based on two reasons: first was that it would standardize American operations, forcing ramp facilities to accommodate their right-sided doors on their Ford Tri-motors; the second reason for right-sided doors was that pilots of the DC-2 always started the left engine before the right, so the DC-3 would also be operated the same way, and by placing the door on the right side, boarding passengers would not be buffeted by the prop wash if the left engine was running and ready for a quick departure.

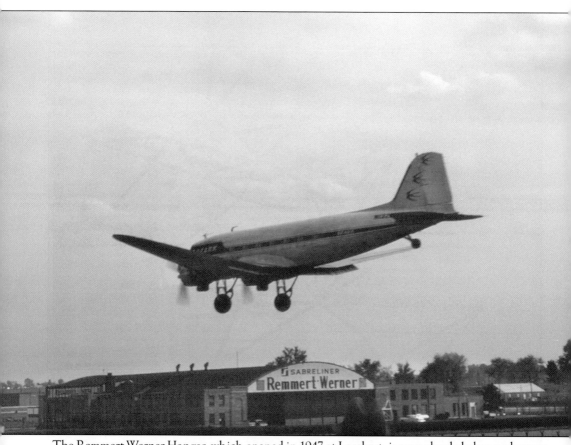

The Remmert-Werner Hangar, which opened in 1947 at Lambert, is seen clearly below and across from this landing Ozark DC-3. Bill Remmert and Bob Werner opened their company with a mission to modify and convert ex-airline/military Douglas DC-3 aircraft, and later the mission included changing Convair aircraft into corporate aircraft. Eventually, as business aviation entered the jet age in the late 1950s, Remmert-Werner became the only completion center in the world for the North American Rockwell Sabreliner aircraft, which was built in El Segundo, California, and then flown to St. Louis to have the interior installed and the aircraft painted. The St. Louis–based company, PET Milk, purchased the first civilian Sabreliner 40, serial number 282-001.

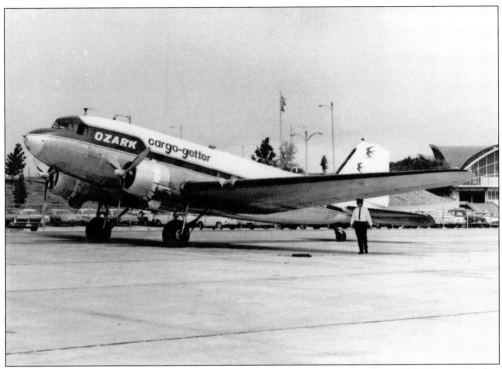

An Ozark cargo DC-3 is being pre-flighted by its captain sometime after 1956. The parabolic, dome-shaped terminal, created by the famed architect Minoru Yamasaki, can be seen in the background to the right.

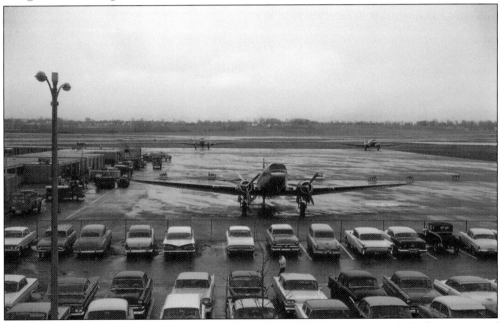

The last scheduled flight of a DC-3 out of Lambert did not occur until the beginning of November 1968. This image was taken at Lambert in 1962 of a Convair 240, which Ozark had added to its fleet.

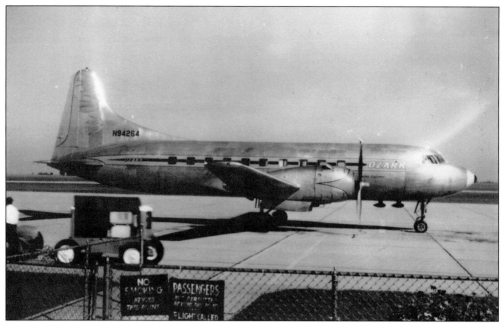

Unlike TWA, which elected the un-pressurized Martin 202 as a replacement for its DC-3 fleet, American Airlines also opted for the Convair, but its fleet was pressurized. At 280 miles per hour, the 202 carried 40 passengers over a distance of 1,200 statute miles. Many of the early Ozark fleet came from retired American aircraft, therefore it was a logical step for Ozark to expand its service with the Convair instead of the Martin 202.

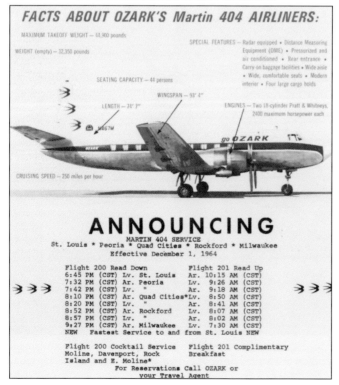

Ozark Airlines did purchase the Martin 404, which had a pressurized cabin. This brochure promotes the features of the aircraft to the passengers of Ozark.

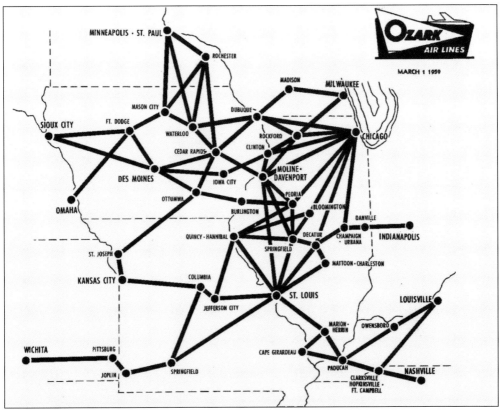

This Ozark Airlines map details routes from 1959.

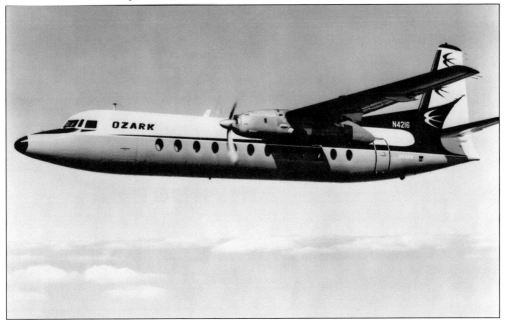

Ozark entered the turbo-prop age when it purchased the Fairchild F-27 and Fairchild Hiller FH-227 aircraft. The FH-227 was a stretch version (longer fuselage and 14 more seats) of the F-27.

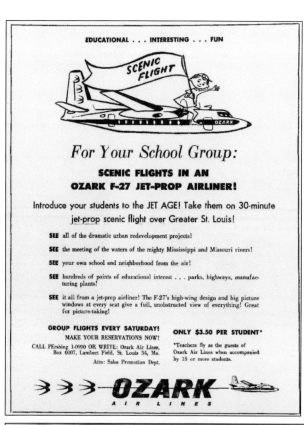

An Ozark Airlines brochure offers scenic flights in an F-27 aircraft. The F-27 series was designed and built by Dutch company Fokker. In the late 1950s, the Fairchild Corporation purchased the license to manufacture the Fokker in the United States. Powered by two 2,250–shaft horsepower Rolls-Royce Dart engines, the F-27 could whisk 44 passengers up to 20,000 feet at 322 miles per hour over a journey of 1,100 statute miles.

An Ozark promotional photograph shows that Ozark was then a very modern airline with both a turbo-propeller and jet aircraft fleet.

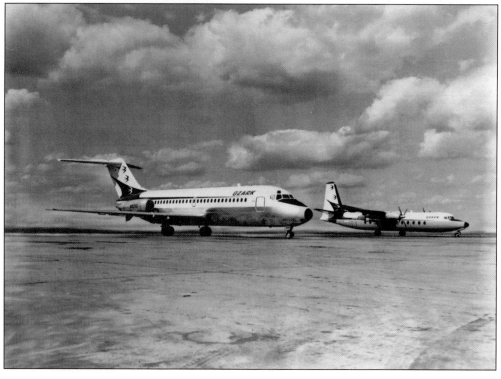

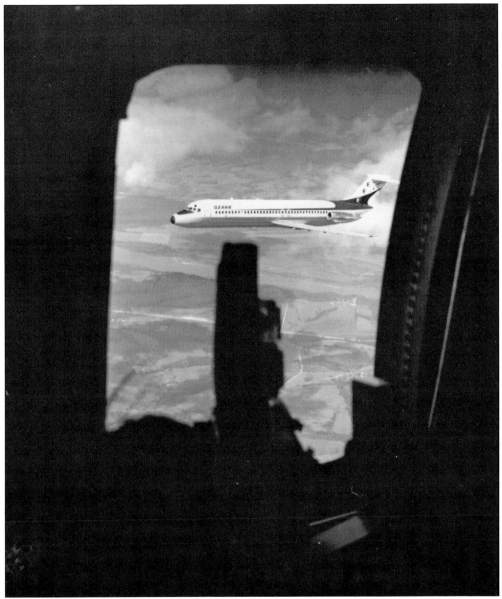

In 1966, Ozark enters the "pure jet age" with service offered in its new Douglas DC-9 aircraft. In this image, the motion picture cameras roll as Ozark's first Douglas DC-9 is flying over the Mississippi River. Footage was used to create promotional films for the airline. The Douglas Commercial DC-9 Series of jet aircraft totally revolutionized the fleet of any airline that chose to purchase it back in 1965. Many smaller airlines like Ozark were accustomed to operating at a sedentary speed range of 200 to 300 miles per hour and flying propeller aircraft, which kept passengers exposed to bumpy weather at 20,000-foot altitudes. The DC-9 was designed specifically to operate from short runways and on short- to medium-range routes, so that the speed, comfort, and reliability of jet transportation could be extended to hundreds of communities previously served only by propeller-driven airliners. The DC-9 provided a cruise speed of 561 miles per hour, cruising altitudes in the upper-30,000-foot range (typically above most thunderstorms), and the capacity to fly 90 passengers. The Pratt and Whitney JT8D-5 Turbofan engines made all of this possible.

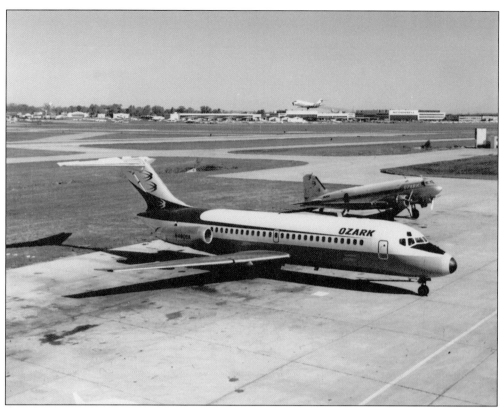

Another interesting image seems incongruous today. The corporate headquarters and factory of McDonnell Aircraft Corporation are clearly visible behind the Ozark DC-3 and DC-9.

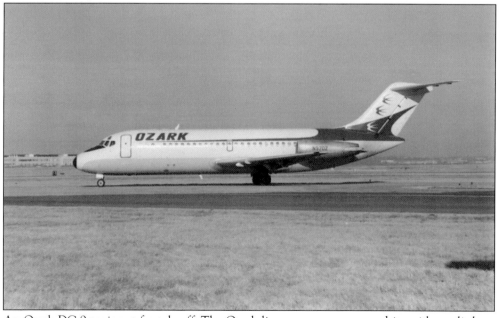

An Ozark DC-9 taxis out for takeoff. The Ozark livery was green over white with a split base of grey.

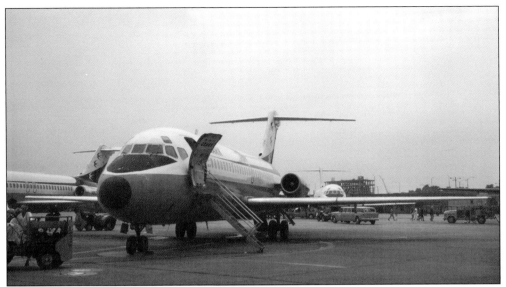

The Ozark ramp at Lambert is busy. Much to management's surprise, Ozark was eventually taken over by its local St. Louis rival, TWA. In 1987, Ozark was purchased for $250 million in a hostile takeover that was led by the new chairman of TWA, Carl Icahn, who had purchased TWA for $350 million, including taking over the company's $1.3 billion of debt, less than a year before the Ozark deal.

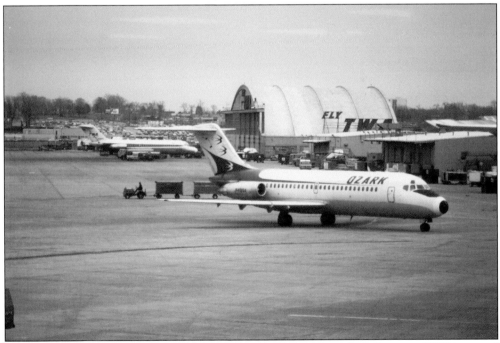

Carl Icahn's acquisition of Ozark, though bad for St. Louis, was a shrewd move—TWA's annual traffic immediately increased by 30 percent, totaling 27 million passengers per year, which brought much needed revenue to clear the debt burdening TWA.

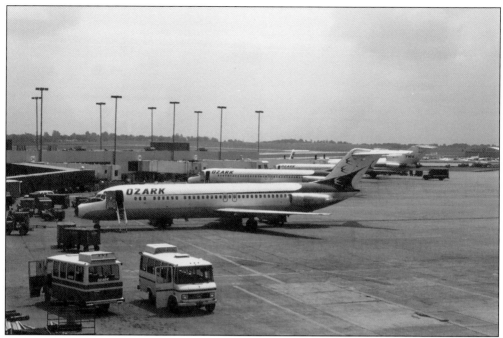

TWA began providing pure-jet service in and out of St. Louis-Lambert Municipal Airport in 1959 with Boeing 707s. This image shows two TWA Boeing 727s and a TWA DC-9 sharing the terminal with as many Ozark DC-9 aircraft.

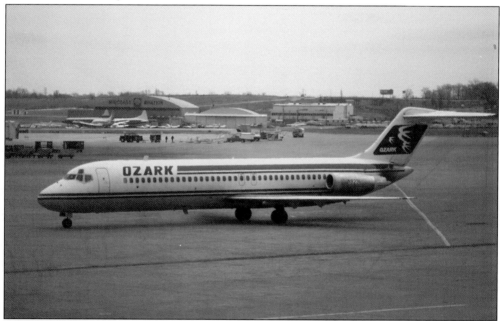

Midcoast Aviation can be seen clearly in the background of this image. Midcoast was formed in 1971, and today it is the world's largest maintenance, repair, and overhaul (MRO) business offering services to the business aviation industry. Midcoast is now owned by Jet Aviation, which is under the control of General Dynamics Corporation. Midcoast left Lambert Field for good in 2009, with all St. Louis operations now based entirely at the St. Louis Downtown Airport.

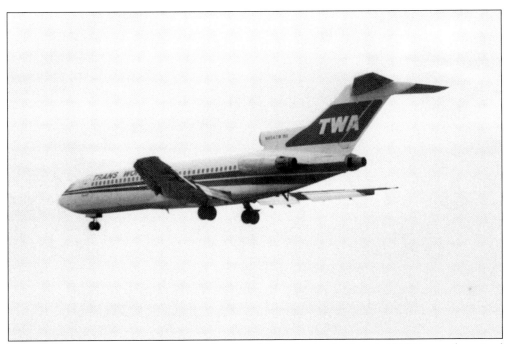

A TWA B-727 approaches Lambert for landing. Until the Anglo-French SST Concorde arrived on the airline scene, the Boeing 727 was the world's fastest airliner at 632 miles per hour.

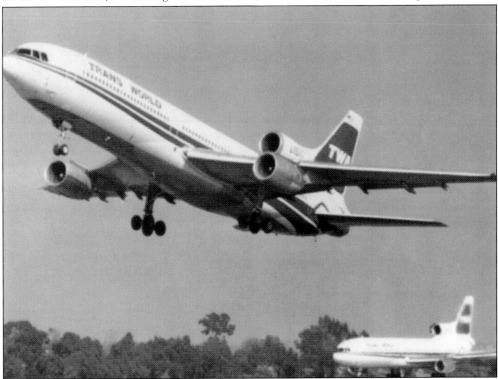

TWA has a long history of maintaining and operating Lockheed products. Here an L-1011 blasts off, while another waits for its turn.

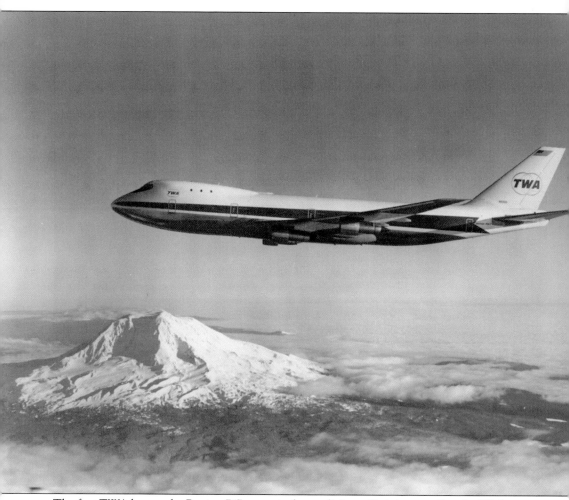

The first TWA heavy, the Boeing 747, poses in front of Mount Rainier near Seattle, Washington, where this fine aircraft was built. The Boeing 747 was the largest aircraft that TWA flew. It was 231 feet long, 196 feet wide, and carried 342 passengers almost as fast as the Boeing 727. There were 58 first-class seats in the front of the TWA B-747 fleet, a 15-place lounge upstairs in the hump behind the cockpit that was reached by a staircase, and 284 coach seats on the main deck. The TWA terminal at New York's Idlewild/Kennedy Airport was the first terminal in the world that was built to serve the massive Boeing 747. TWA's inaugural service of the B-747, which was the first in the industry, began on February 25, 1970.

Five

Missouri Air National Guard and St. Louis Naval Air Station

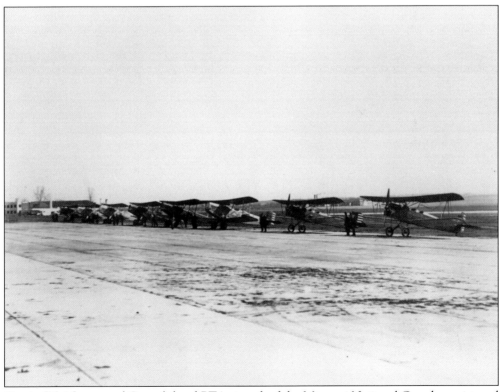

Curtiss JN-4 Jenny and Consolidated PT-1 aircraft of the Missouri National Guard are pictured on the ramp at Lambert around 1927.

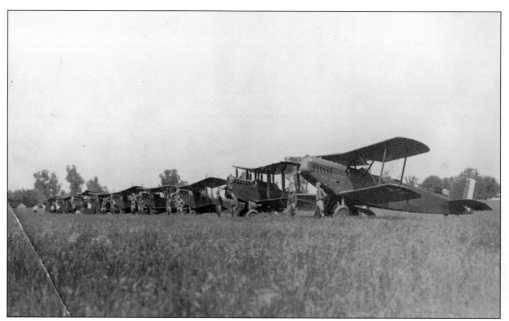

The 110th Observation Squadron aircraft maneuver at Lambert Field in 1924.

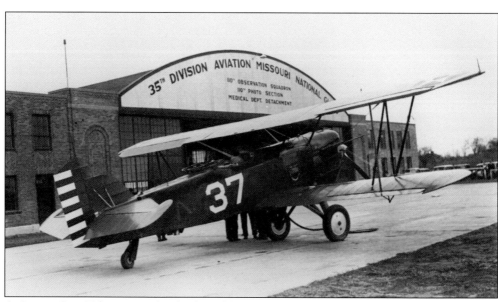

A Curtiss O-11 Falcon aircraft, which could reach up to 110 miles per hour, readies for a sortie from the squadron hanger.

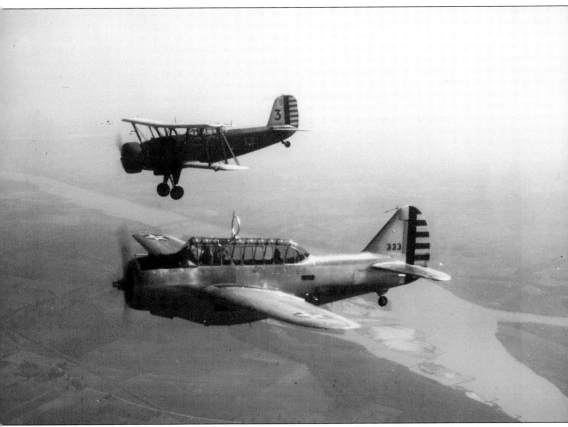

A North American O-47A Guard monoplane (foreground), with the capability of reaching 200 miles per hour, and a Douglas O-38E Guard biplane, which saw speeds as high as 130 miles per hour, are seen flying over the Mississippi River. The ring above the cockpit of the O-47 is a loop-sense antenna used to point toward a radio transmission to determine the signal's direction.

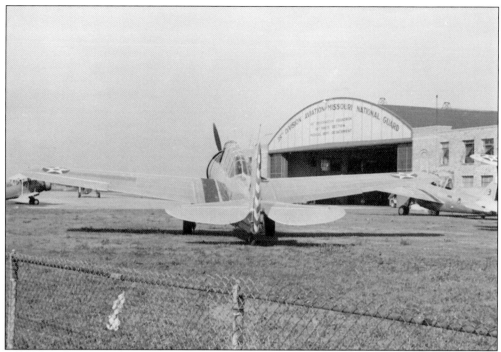

Various 110th Observation aircraft are pictured, including a rear view of a North American O-47A aircraft.

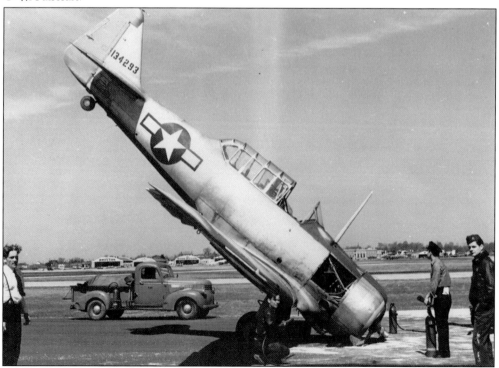

An AT-6D Texan aircraft tipped onto its nose while taxiing too quickly at Lambert on a visit to the 110th Observation Squadron.

A F-51D gets an engine change in the field. The North American P-51D was powered by a liquid-cooled Packard V-1650, two-stage geared-supercharged V-12 piston engine that produced 1,720 horsepower. The P-51D had a 437-miles-per-hour maximum-level flight speed and would normally cruise at 362 miles per hour.

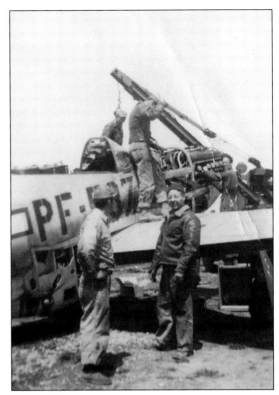

Here are more F-51Ds from Missouri during their deployment in Asia while serving in the Korean War. The Mustang on the left displays the Missouri National Guard letters, MONG.

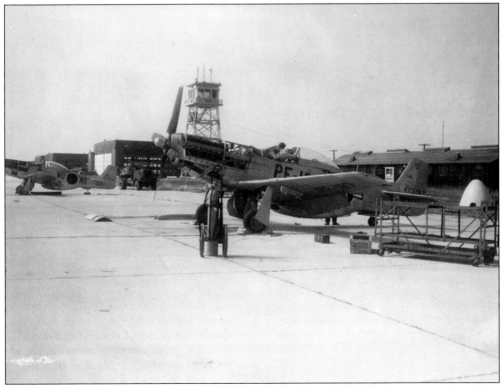

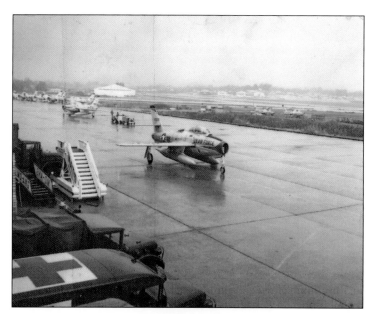

Seen in October 1961, F-84F Thunderstreaks are leave home base and are bound for Toul-Rosleres Airbase in France in response to the escalating situation in Berlin. Nikita Khrushchev had ordered that all Western forces leave West Berlin. Cold War tension had escalated to a fevered pitch after Gary Powers had been shot down 18 months before. The deployment in France ended on their return in August 1962.

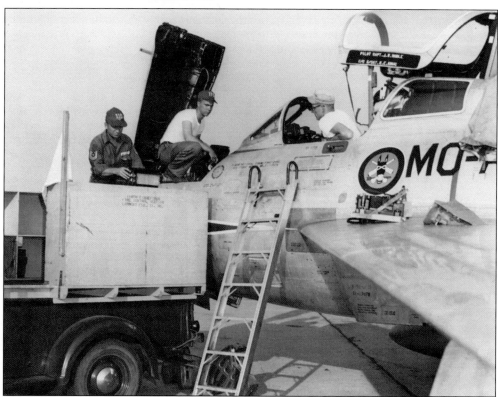

MOANG avionics technicians get under the hood of an F-84F to fix an avionics problem with this aircraft at Lambert. The Republic F-84F first arrived at the Missouri Air National Guard in 1957 (the year that the wing received its first jet aircraft). It was supplied with the Lockheed F-80 Shooting Star and also the Lockheed T-33 (T-Bird) as a training aircraft. Then, in the summer of 1957, the F-84F arrived.

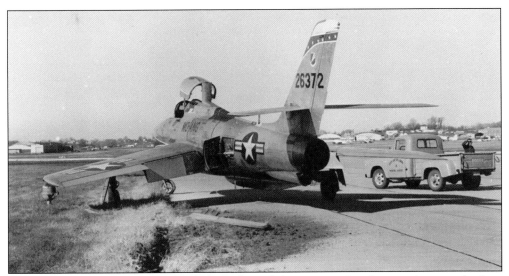

Here is a plane that encountered a slight mishap while taxiing at Lambert; nothing was broken, except maybe the pilot's ego. The St. Louis wing flew here until June 13, 2009, when the Missouri Air National Guard left Lambert-St. Louis International Airport, ending more than 86 years of military operational history at this venerable airport.

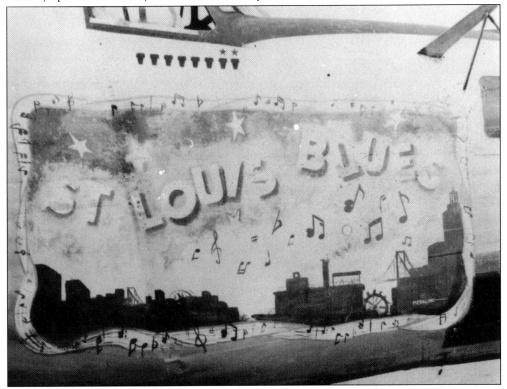

St. Louis Blues was a F-7A / B-24J-5-CF Liberator, serial number 42-64172, from the 20th Combat Mapping Squadron, 6th Photo Recon Group, 5th Air Force. This aircraft flew 16 missions during World War II, and during its operational career it was based in Colorado Springs and Manila in the Philippines. It came through St. Louis on a training exercise in early 1943.

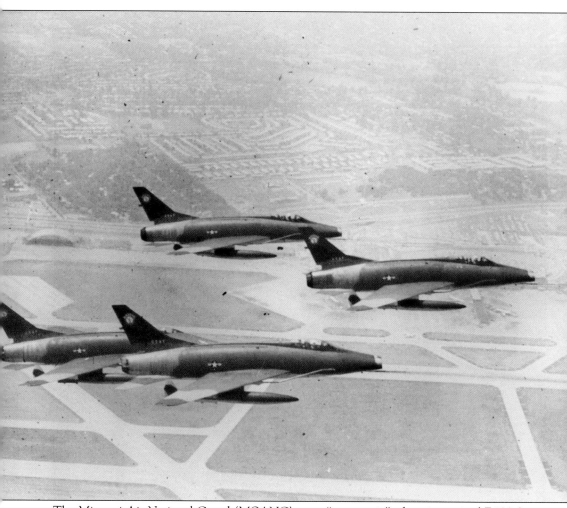

The Missouri Air National Guard (MOANG) went "supersonic" when it received F-100 Super Sabre aircraft in late 1962. MOANG F-100Cs are pictured over here Lambert. The Super Sabre was the first fighter-jet aircraft to be able to reach supersonic speed in level flight. This incredible capability was enabled by the single Pratt and Whitney J57-P-21 turbojet engine that produced 16,000 pounds of raw thrust when the afterburner was activated, which is when raw jet fuel is sprayed into the jet pipe after body and the exhaust nozzle opening is increased at the same time. Without use of the afterburner, the engine produced 10,200 pounds of thrust. When the afterburners were utilized, it was possible to see the fuel gauges move in the cockpit. Its fuel consumption at that time was about 20,000 pounds per hour, which is a little over 2,800 U.S. gallons per hour, and the aircraft only carried 2,139 U.S. gallons, including its drop tanks. The ceiling of the F-100 was 51,000 feet, and the empty weight was 19,270 pounds.

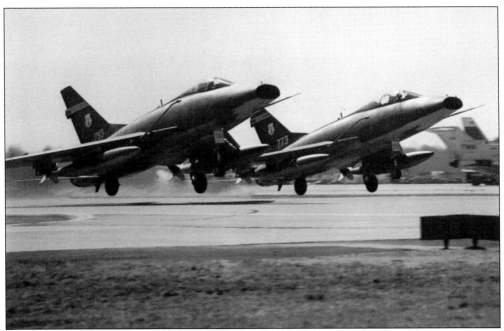

The North American F-100 Super Sabre remained in service with the MOANG for the next 16 years, until it converted over to F-4C Phantom II aircraft, which was built across the runway from its base at McDonnell Aircraft Corporation at Lambert Field. The last MOANG F-100 was flown operationally in 1965.

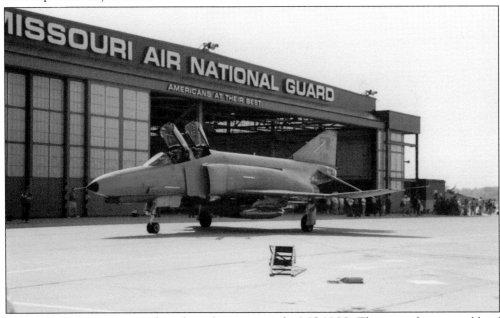

The F-4E Phantom was another plane that came to the MOANG. This aircraft was capable of exceeding twice the speed of sound—double supersonic. The F-4 Phantom II was built in St. Louis at Lambert Field between 1958 through 1979. A total of 5,195 F-4 Phantom II aircraft were built. Of those, 5,057 were produced at Lambert, while the last 138 were built under license by Mitsubishi Aircraft Company in Japan.

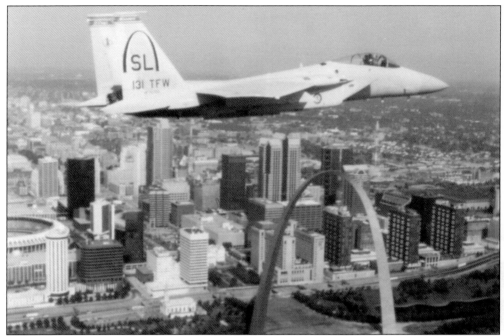

A MOANG F-15 poses in front of the Arch, as it slides by this architectural wonder. The F-15 Eagle is also a McDonnell Douglas (now Boeing) product that is built at Lambert Field. The development of this aircraft was in direct response to the cold War Russian MIG-25 Foxbat. The contract to build the F-15 was awarded to McDonnell Douglas in late 1969. Its first flight was in 1972; the MOANG received its first F-15 in 1991.

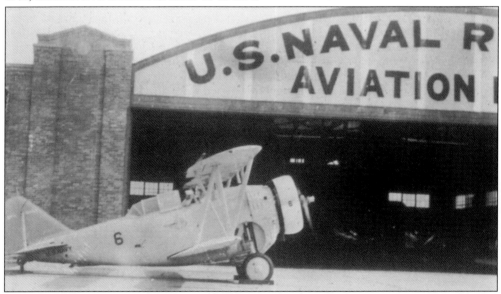

A Naval Reserve Curtiss BF2-C1 (Model 67A) Goshawk fighter/bomber with retractable undercarriage sits on the ramp. In 1925, two years after the 110th Observation Squadron of the Missouri Air National Guard made Lambert its home base, the Naval Reserve Unit of the St. Louis Naval Reserve Armory was formed at Lambert Airfield. It would later officially become the Naval Air Station of St. Louis.

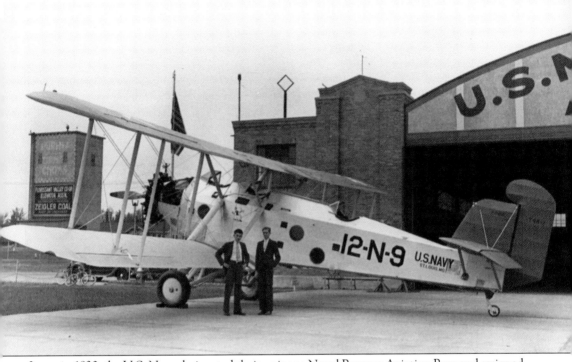

Later, in 1930, the U.S. Navy designated their unit as a Naval Reserve Aviation Base and assigned Lt. Frank Weld as the base commanding officer. Immediately thereafter, two N2C2 Curtiss training aircraft and three O2C-1 Curtiss Helldiver aircraft were delivered to the new base at Lambert. The City Of St. Louis built the navy a hangar on the northwest corner of the airport, which was used until 1942, when the navy built a much larger air station base on the south side of the airport. This facility was turned over to the Missouri Air National Guard in 1958, when the station was closed. In this photograph is a naval reserve Martin T4M-1 torpedo bomber aircraft parked outside of the base hangar. This aircraft is unusual because it has three separate cockpits, one for each crew member. It is equipped with a 525-horsepower Pratt and Whitney Hornet engine.

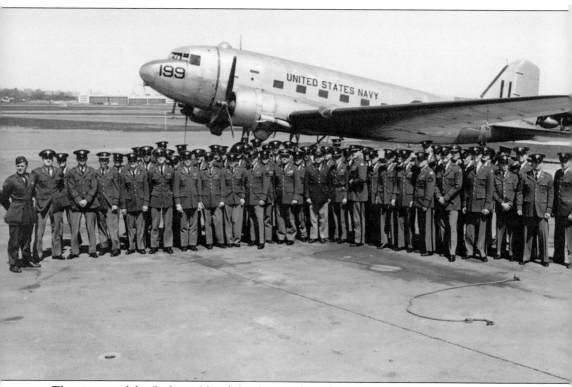

The aircrew of the St. Louis Naval Air Station (NAS) poses here in front of a Douglas C-47 at the end of World War II. Notice the Perspex dome on top of the fuselage of the C-47, just behind the cockpit. This is the celestial navigation dome through which the navigator would use his sextant to determine his aircraft's position and to plot it on his aviation chart. As a side note, the "Dakota" name was coined by the British forces in World War II, when they put "Douglas Aircraft Company Transport Aircraft" together to form "DACoTA, hence "Dakota!"

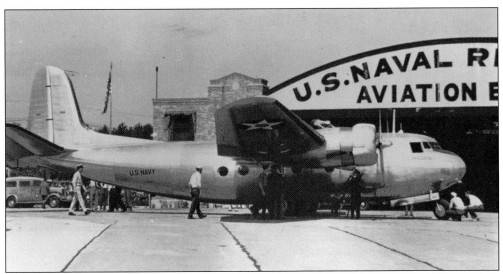

An R3D-1 Naval Air Transport aircraft is shown at the St. Louis NAS in 1940. Though in civilian guise, this Douglas aircraft is technically a DC-5. A comparison of the nose section and the fuselage, located just aft of the wing trailing edge, shows a striking resemblance to the DC-4.

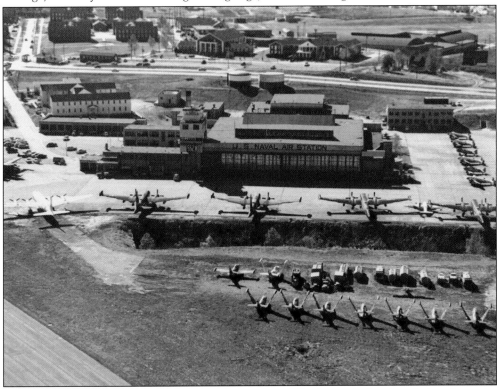

An aerial view shows the St. Louis NAS at Lambert. The large, dark-colored aircraft on the concrete ramp are Lockheed P2V-5 Neptune aircraft. The white-looking aircraft to the far left is a Douglas DC-4, designated by the navy as a R5-D Transport. To the right is the station's fleet of Beechcraft SNB-5 aircraft (the civilian designation is BE18). In the foreground, McDonnell F2H-2 aircraft carrier jet fighters sit on the grass.

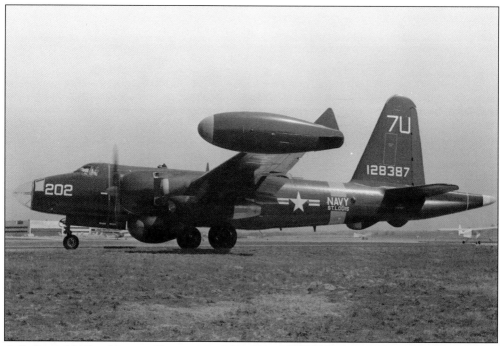

A St. Louis NAS radar surveillance aircraft Lockheed P2V-5 Neptune taxis out to the runway.

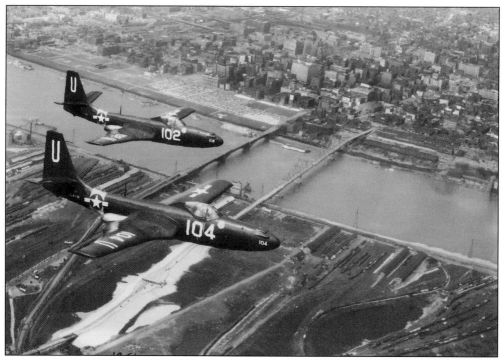

St. Louis–built McDonnell FH-1 Phantom I aircraft of the St. Louis NAS fly past downtown St. Louis in 1951.

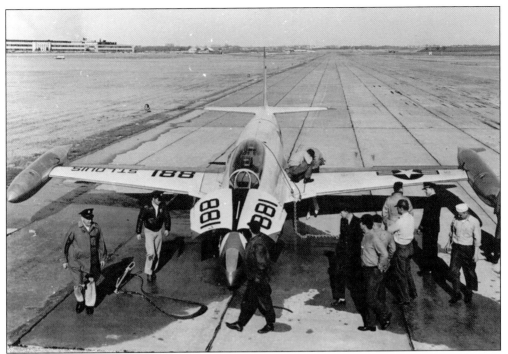

One of the two Lockheed TV-2 (T-33 Shooting Star) training aircraft that was stationed in St. Louis NAS is prepared for towing after a failure of its nose wheel.

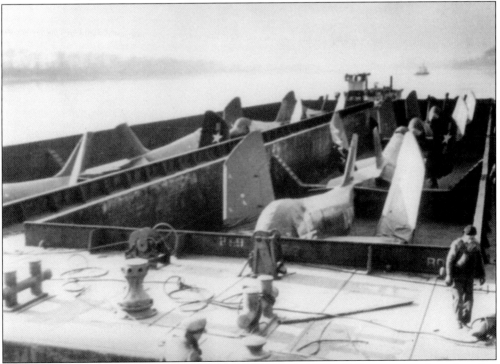

The St. Louis NAS was decommissioned on February 1, 1958; all flying stopped on that day. F3H-1 Demons are shown on their way to Memphis via a Mississippi barge.

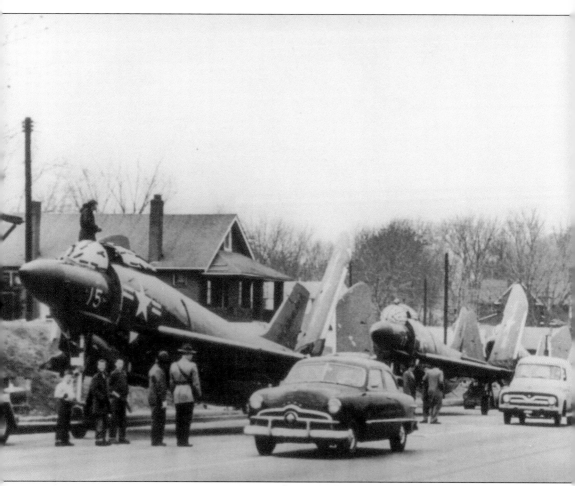

The St. Louis NAS was closed by 1959, and the facilities were handed over to the MOANG. Remaining aircraft—which either could not be flown or for which there were no available pilots—were transported by road and then barge down the Mississippi to the Memphis NAS. Here are F3H-1 Demons being towed to the barge terminal for loading and shipping.

Six

AIRCRAFT MANUFACTURING AND THE SPACE RACE

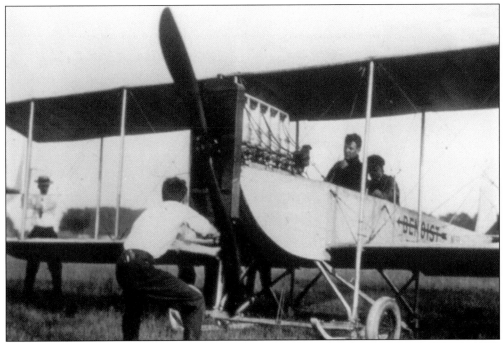

A Benoist Tractor biplane is pulled through and ready for a "hand-prop" start. In March 1908, Thomas W. Benoist established the nation's first aeronautical supply company, christened Aerosco, at 3932 Olive Street in St. Louis.

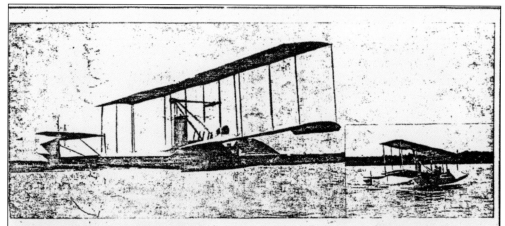

The Benoist Airboat
represents the last word in Hydroaeroplane construction. The only flying boat in America built with the motor down in the hull, this giving perfect stability when on the water with a maximum of flying stability. Carries two passengers with ease. *Write for particulars.*

BENOIST AIR CRAFT CO.
UNIVERSITY CITY   ·   ·   -   MISSOURI

An early advertisement touts the Thomas Benoist Flying Boat Company. Benoist was the first native of St. Louis to pilot an aircraft and also the first to use the Kinloch Airfield in September 1910. In 1911, Benoist opened St. Louis's first flying school. He then turned his attention to the aircraft design and manufacturing business.

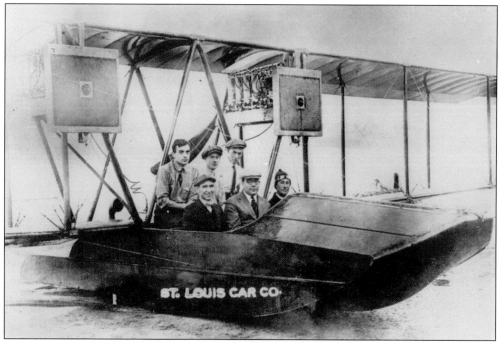

A Benoist twin-engine design Flying Boat is pictured on the Mississippi. Seen from left to right in the front row are Thomas Benoist, Edwin Meissner of the St. Louis Car Company, and Tony Jannus in the pilot seat.

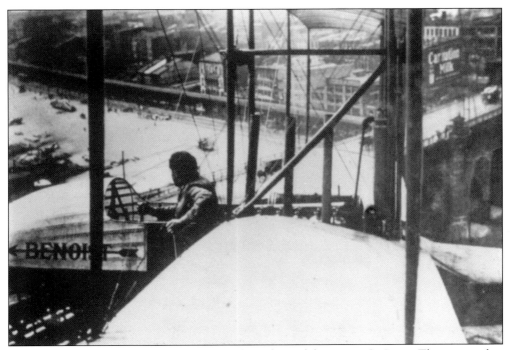

A Benoist Flying Boat is pictured over the Eads Bridge and downtown St. Louis. The camera that captured this image was mounted on the outboard interplane struts of the left wing, and the pilot activated its shutter by pulling a string, which can be seen in the photograph.

This image of a two-seat Benoist Flying Boat was taken in 1912. In September 1917, the St. Louis Aircraft Corporation was formed, mainly thanks to the expertise that Thomas Benoist had developed in manufacturing his own aircraft designs.

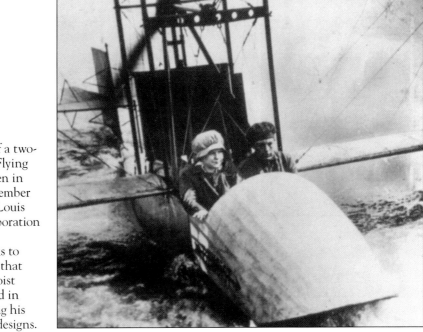

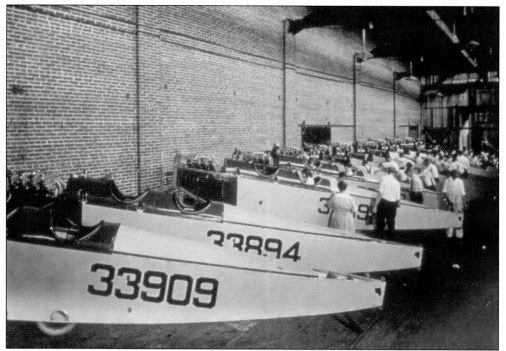

The St. Louis Aircraft Corporation was given a commercial order by the U.S. War Department to produce Curtiss JN-4 Jenny aircraft for the Joint Army and Navy Technical Aircraft Board of the U.S. government to support the United States and Allied forces fighting against Imperial German armies in France and Belgium in World War I. This image shows JN-4s during construction; 450 Jennys were produced before Armistice was declared on November 11, 1918.

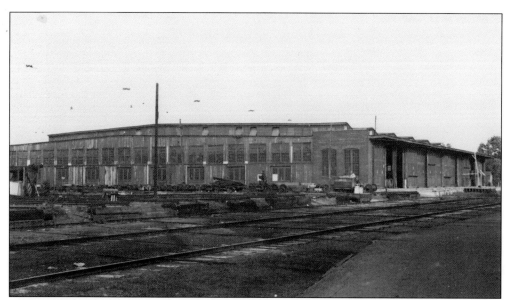

The parent company of St. Louis Aircraft Corporation was the St. Louis Car Company. This image shows one of the car company's manufacturing plants that were ceded over to the aircraft company under emergency orders.

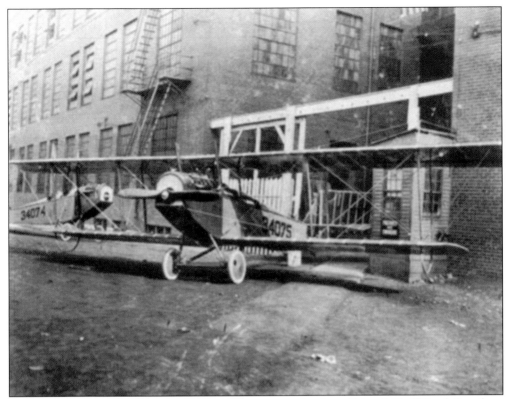

Completed Jennys at the plant wait testing at the company's private flying field off North Broadway in North St. Louis.

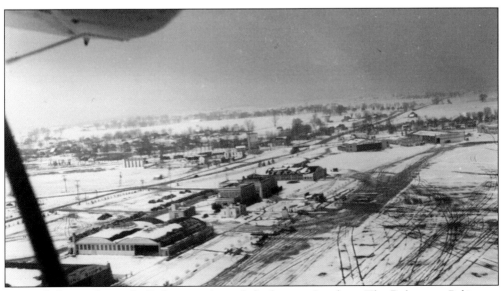

An angled view shows the Robertson Aircraft facilities at Lambert. The Robertson *Robin* was built here.

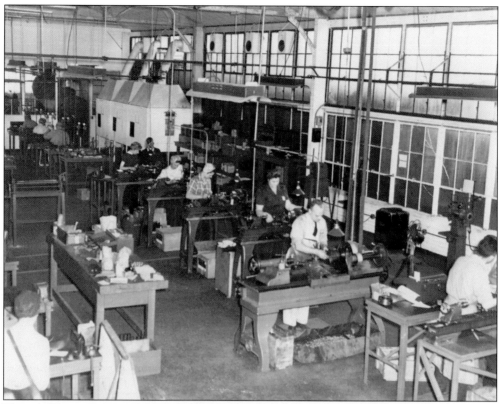

The machining section of the Robertson Aircraft Plant is pictured above. The fabric covering section of the Robertson Aircraft Plant is pictured below.

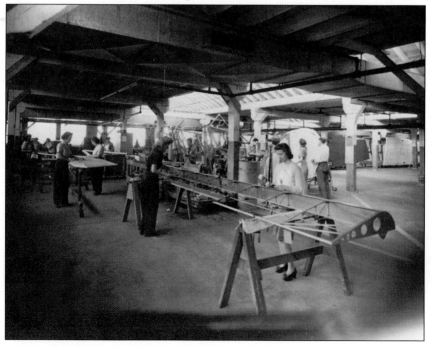

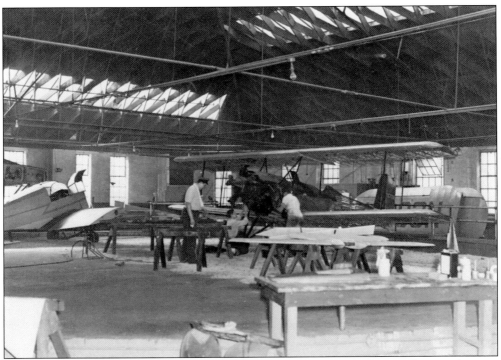

The production fitting section employees of the Robertson Aircraft Plant are busy working on a plane.

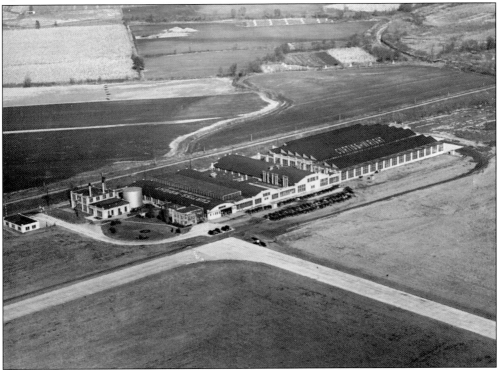

The Curtiss-Wright Plant sat on the north side of Lambert Field.

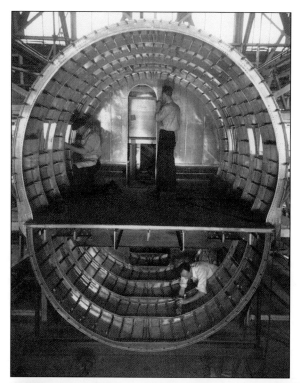

At left, the Curtiss CW-20 Condor airliner is pictured under construction at the Curtiss Plant at Lambert Field. The Condor production line is pictured below.

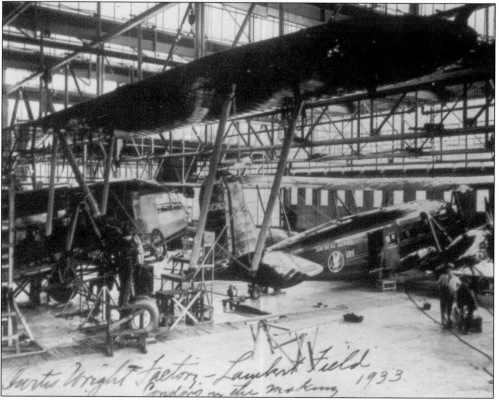

Curtis Wright Factory - Lambert Field 1933
Condors in the making

114

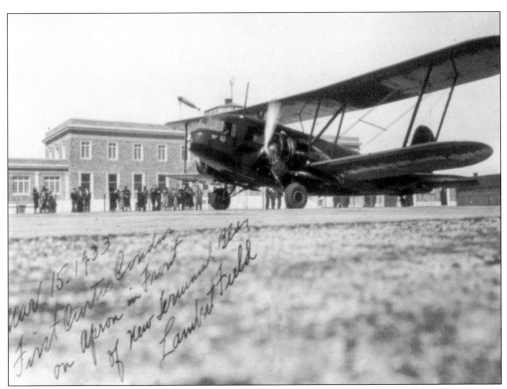

March 15-1933
First Curtiss Condor
First Condor on Apron in Front
of New Terminal Bldg
Lambert Field

The first Condor came off the production line in 1933. The Lambert Terminal building can be seen in the background in the photograph above. Before it was delivered to American Airlines, a Condor is pictured below on its first flight along the Missouri River.

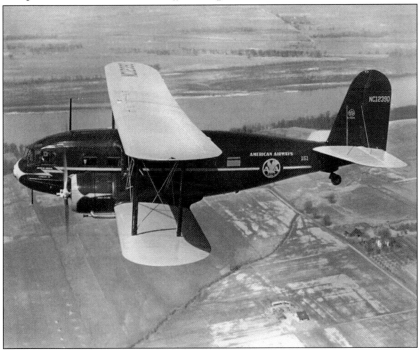

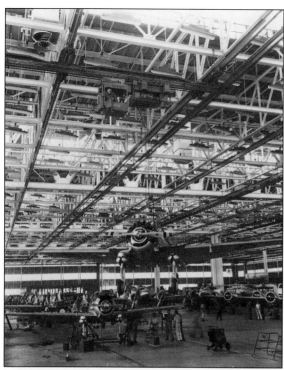

The SNC and AT-9 production line at Curtiss-Wright on Lambert Field is pictured at left. Wartime production of the SB2C Helldiver is pictured below.

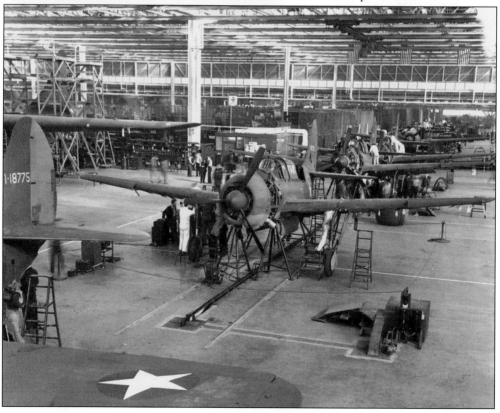

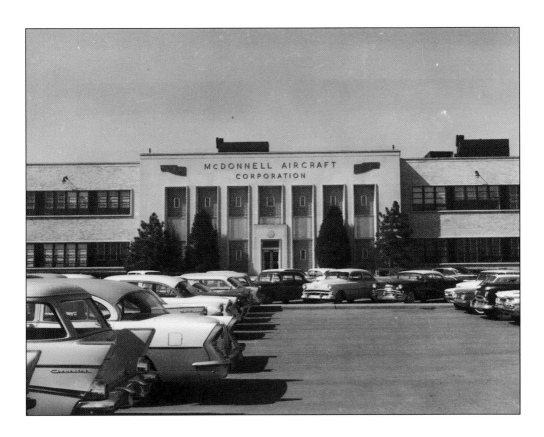

In the summer of 1939, James Smith McDonnell Jr. founded the McDonnell Aircraft Corporation at St. Louis-Lambert Municipal Airport. The above image of corporate headquarters was taken in the mid-1950s. Below is an aerial view of the McDonnell Aircraft Plant, taken in October 1955.

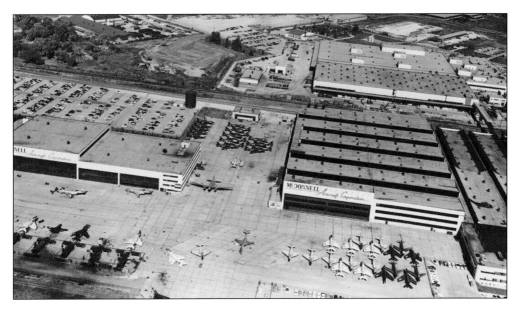

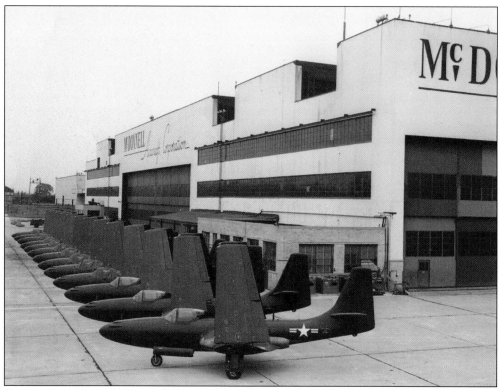

On July 21, 1946, the McDonnell FH-1 Phantom became the world's first jet aircraft to takeoff and land on an aircraft carrier, the USS *Franklin D. Roosevelt*. This image shows an entire squadron of Phantoms fresh off the assembly line in St. Louis.

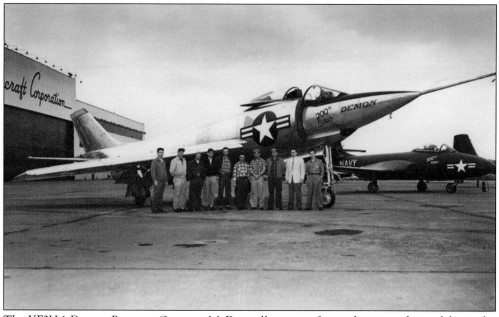

The XF3H-1 Demon Program Group at McDonnell poses in front of its aircraft to celebrate the 200th flight of this machine.

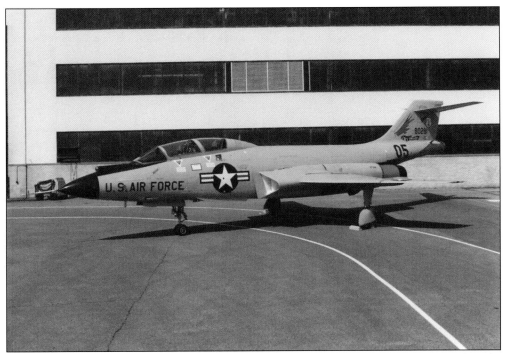

Here is an example of the F-101 version of the Voodoo that was produced in large numbers by the McDonnell Aircraft Corporation.

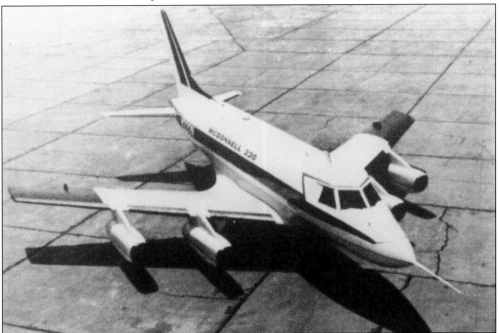

In its bid to win the Air force Utility Transport Experimental (UTX) program, the McDonnell 119/220 was designed and produced. Unfortunately this design lost to the Lockheed JetStar, and McDonnell cancelled the program. The prototype was flown for almost 30 years as a corporate aircraft for the company until it was sold into private hands.

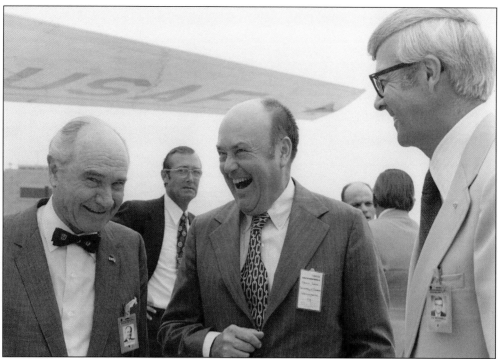

During an official visit to the McDonnell plant in 1972, Secretary of Defense Melvin Laird (left) is seen joking with James (center) and Sanford McDonnell.

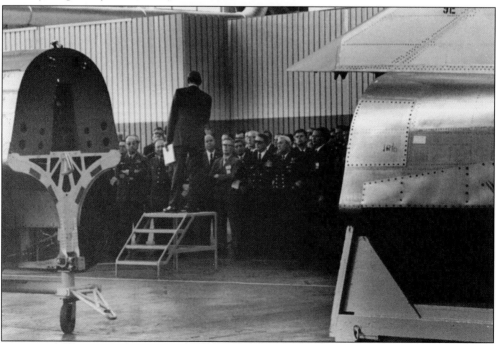

In 1961, the McDonnell F4H Phantom II became the world's fastest aircraft when it was recorded at a speed of 1,606.3 miles per hour. Air force officials took this image during a visit to the Phantom II production line in St. Louis.

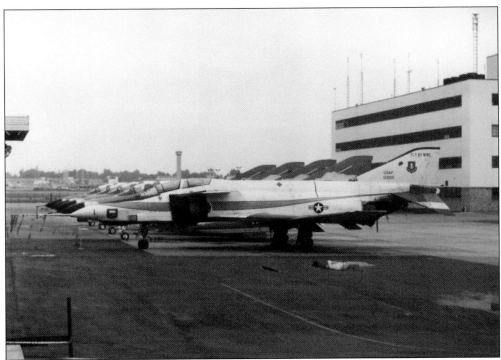

Above, four standard Phantom IIs are parked alongside a "flyby-wire" test ship on the McDonnell Flight Test Ramp. Below, an F-4B Phantom II retracts its gear just after breaking ground.

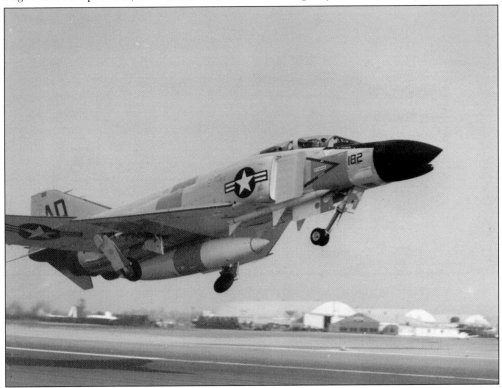

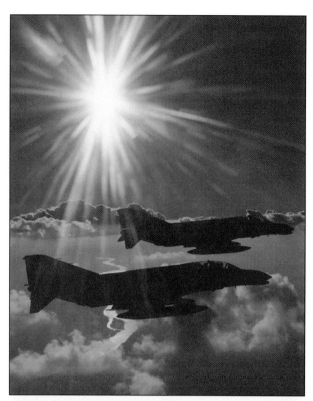

F-4C Phantom II aircraft of the MOANG group at Lambert Field are shown at left. Below, Joe Dobronski, the chief test pilot for McDonnell, signed this photograph after he made the first flight in the Blue Angels version of the Phantom II, the F-4J model.

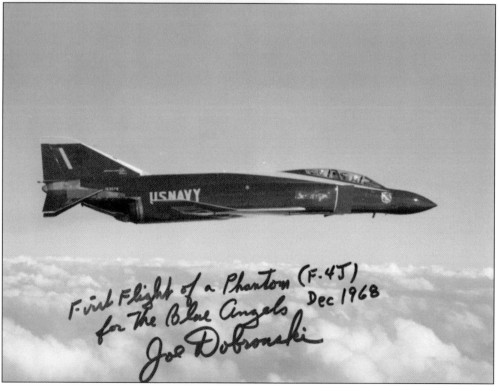

First Flight of a Phantom (F-4J) for The Blue Angels Dec 1968 Joe Dobronski

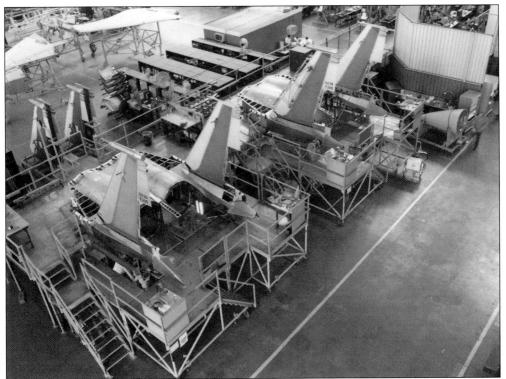

The F-4 Phantom II and the F-15 Eagle shared the same production line until the F-4 program reached its end in 1981. F-15 Eagle aft fuselages are seen being built and held in their jigs until they are mated with the forward section and wings. Jim Lentz, who is now retired from McDonnell Douglas, set up the promotional shot below.

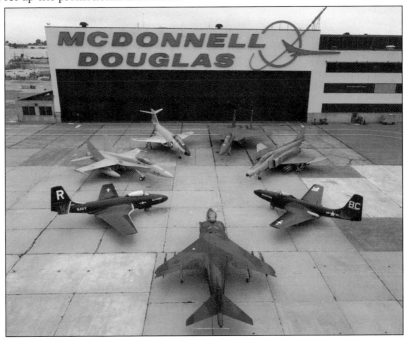

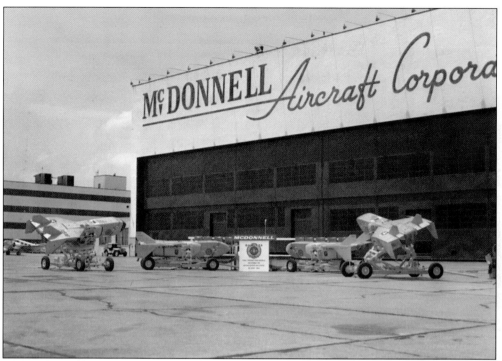

McDonnell, a producer of missiles and other "pilotless" aircraft, would later merge with the Douglas Company. McDonnell also became well known for its groundbreaking helicopter designs.

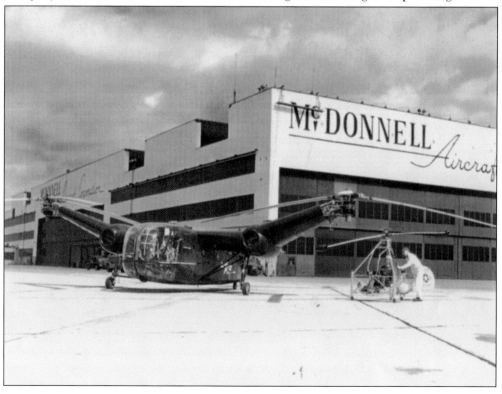

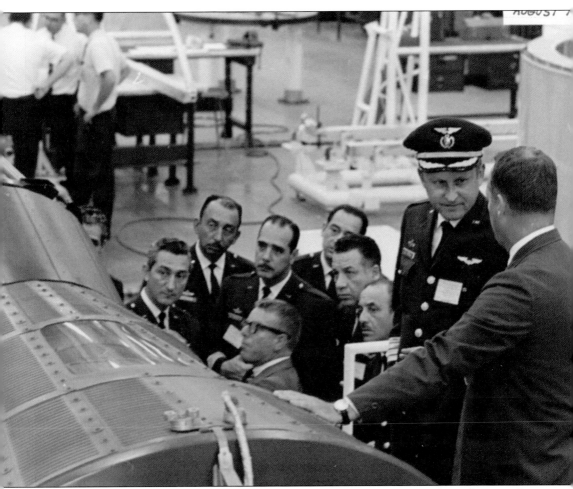

Military top brass inspect progress with the McDonnell Mercury Capsule Program. McDonnell Aircraft contributed to the NASA Mercury and Gemini space programs.

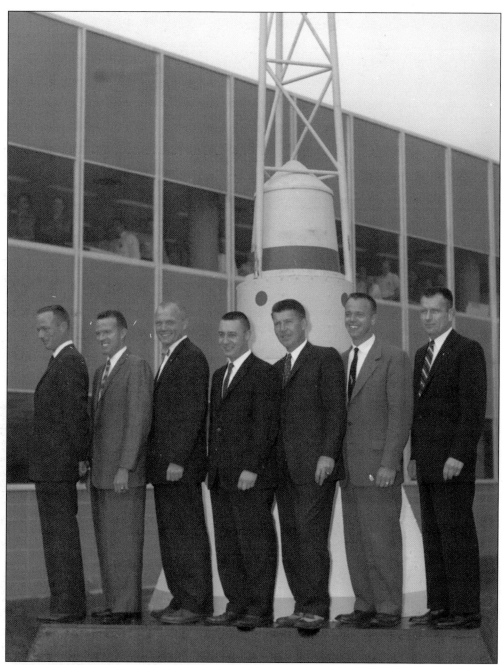

The Gemini astronauts are photographed at McDonnell Aircraft in 1963. McDonnell Aircraft made Pres. John F. Kennedy's promises, which were declared in the famous speech at Rice University, possible. Kennedy said, "We choose to go to the moon. We choose to go to the moon in this decade and do the other things, not because they are easy, but because they are hard, because that goal will serve to organize and measure the best of our energies and skills, because that challenge is one that we are willing to accept, one we are unwilling to postpone, and one which we intend to win."

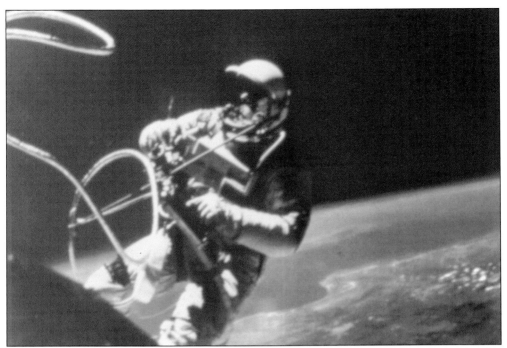

Astronaut Ed White became the first American to "space walk," or to engage in extra-vehicular activity (EVA), as termed at NASA, on June 3, 1965, via a McDonnell Aircraft Corporation Gemini Capsule.

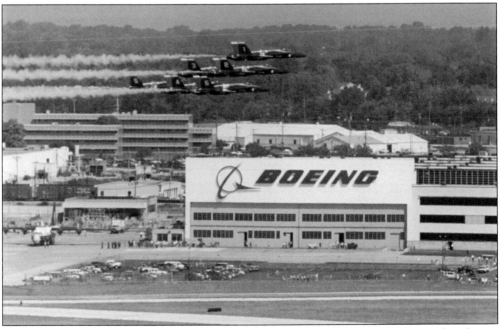

On August 1, 1997, the McDonnell Douglas Corporation ceased to exist after it was bought out by and merged into the Boeing Company. Since 1908, a total of 108 differing models of aircraft, gliders, airships, and helicopters have all been designed and built in St. Louis. Tens of thousands have been, and are still, produced to this day.

DISCOVER THOUSANDS OF LOCAL HISTORY BOOKS FEATURING MILLIONS OF VINTAGE IMAGES

Arcadia Publishing, the leading local history publisher in the United States, is committed to making history accessible and meaningful through publishing books that celebrate and preserve the heritage of America's people and places.

Find more books like this at
www.arcadiapublishing.com

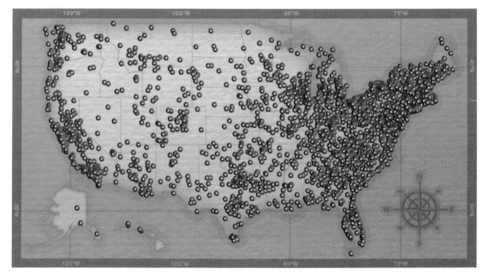

Search for your hometown history, your old stomping grounds, and even your favorite sports team.

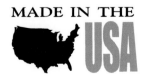